Cosmology of God

and

The Universe

Gerald Grushow

i

Cosmology of God and the Universe by Gerald Grushow

Copyright © 2015 by Gerald Grushow

LCCN: 2015906845

CreateSpace Independent Publishing Platform, North Charleston, SC

Printed by Create Space, An Amazon.com Company

ISBN-13: 978-1511847728

ISBN-10: 1511847727

Available from Amazon.com, CreateSpace.com, and other retail outlets

Books by Gerald Grushow

The Natural God of Law, Love, and Truth 1994
Doppler Space Time 2000
Science of God 2001
Aliens Within Us 2005
Futureoids 2013
Cosmic Reincarnation 2015
These books are available on Amazon.com and other retail
outlets.

Biography

Born Dec. 24, 1938   Brooklyn N.Y.
Brooklyn Tech H.S. Class of 1952
Polytechnic Institute of Brooklyn N.Y. 1956-1966
B.S.E.E. (summa cum laude)
Con Edison Engineering Design Assistant 1956-1962
City of N.Y. Associate Engineer 1962-1967
Sperry Rand Assistant Engineer 1967-1970
Port of NY Authority Engineer 1970-1972
Sperry Rand Radar Research Engineer 1972- 1993
Retired in 1993 to work full time on Theories of God and
the Universe which began in 1981.

With my greatest appreciation to my daughter Melissa who has helped with the spiritual concepts, editing, proof reading, and the book cover. Without her valuable ideas and help, this book would not have been possible.

Preface

This book is an effort to present an understanding of God and the Universe from philosophical and scientific viewpoints. There are few people in the world who could comprehend the mathematical complexity of the multi dimensional multi light speed universe in which we live. As an Engineer I rely upon simple models and simple algebraic equations to understand the universe that we live in.

Over the years my dreams and visions have provided me with scientific concepts which help me to understand the complexity of God and the Universe. To this data, I combine the latest scientific concepts of Einsteinian space time, quantum Mechanics, and String theory together with my Dot-wave theory, and my Doppler space time to arrive at a solution from an Engineering perspective.

The actual scientific solution is a goal for future mathematicians and scientists. The problem is very difficult because we exist in the mind and body of God. All we can do is produce theories and equations which agree with our instruments and measurements. However we can only see and measure a small part of the totality of God and the universe. This gives us a very difficult task to accomplish. All that I can do is to provide some insights into that final goal.

To that effort the Dot-wave theory is designed to look at the material world such that God's reality is an inverted image of our reality. The dot-wave equations and theory is an attempt to produce an engineering model of the universe as viewed from the eyes of God.

Table of contents

Introduction

I have been working on an understanding of God and the Universe for thirty three years ever since a strong and very demanding spiritual encounter occurred in the spring of 1981. In order to understand my lifetime of spiritual encounters, I have studied many possible theories in an attempt to combine physics, philosophy, religion, cosmic reincarnation theory, and my spiritual encounters into a pluralistic theory of God and the Universe.

I was born December 24, 1938 to a Jewish mother and at the time of the year when both the Hanukah and Christmas holidays occur. From the time I was a little child I always had audio communication with the spiritual world.

I have a visionary mind which enables me to readily communicate with my spiritual dimension. This occurs in both audio and audio/visual dreams. It also occurs when I push myself to an excited state as I try to communicate with the spiritual world. This brings me into a visionary state in which new information flows into my mind.

I wanted to understand about God and the Universe. In the year of 198 I brought my mind up to an excited state in order to solve the Rubik cube. I did this only after I had tried to solve it for many hours. Once I was in the excited state, I solved the cube in a few seconds. The excited state gave me ready communication between my spiritual dimension, which is my soul, and my physical mind.

Over the years I have come to realize that most of the data I received did not come directly from God but came from a lower spiritual dimension and from there to my physical brain. It was only during the last few months that I have come to understand that my soul exists within the lower mind of God and contains a superior spiritual brain. I understand now that our soul really belongs to God and is part of God's spiritual energy.

As I reached my seventy-sixth birthday, new data started to come in. It came to me in audio dreams. I would awaken each day and type out the concepts presented to my physical mind from my spiritual mind. I thought that these communications were finished years ago. I thought that I was free of these kinds of dreams and visions. I made no attempt to bring myself up to an excited state. Everything was quiet and peaceful and I was content to leave it be.

For years I have always taken the position of half believer and half atheist. The atheist within me discounts the data as mere dreams. The believer in me always says that the data comes from God. The atheist gives me comfort as it reduces me to an ordinary person with no responsibilities. It also alleviates the great fear of having to face the spiritual world after I am dead. In fact the atheist within me does not believe in the spiritual world at all.

The big problem that I have had is with my soul. It is like that I have an ancient Rabbi living within me. At this writing I have come to realize that the ancient Rabbi is my soul and exists within the lower mind of God. This leads me to the scary conclusion that I am a cosmic reincarnate.

The visions of 1981 specified that I could not die. This gave me great grief. I have had to accept that I have a multiple life soul. To me this was a sad and hopeless condition. That year changed my life and put me on a path to understand my condition and the condition of all of us.

Instead of being a brilliant Electrical Engineer, I was reduced to being a student of God and the Universe. My soul told me that I was ignorant and demanded that I work and study. This I have done to the best of my ability. At different times I would get audio and audio/visual dreams which flowed from my spiritual mind into my physical mind. Then I would start to write anew.

In spite of my writings, I never reached the understanding that I sought. My physical mind was always in conflict with my spiritual mind. As I write my various books, I try to understand God and the Universe from different perspectives. I try to look at God from a Jewish perspective, a Judeo/Christian perspective, and a pluralistic perspective. This book is written from a multi religious perspective so that people of all faiths can gain insights into God and the Universe from a general religious perspective.

My physics book, Doppler Space Time with the Dot theory was self-published in December 2000 and provides some scientific insights into the multi-light speed universe. As I have studied the latest scientific theories, I have added the multi-dimensional Universe to the multi-light speed universe. This produces a much more complex universe which was the basis of my novel "Futureoids" which I self-published in 2013. This philosophical sci-fi novel enables the reader to get a good picture of the total universe in which we live. The novel explains the very complex structure of God from a general religious perspective.

In "Science of God", which I self-published in 2001, I attempted to look at the solution to God and the Universe from a strictly Jewish viewpoint. In the book my encounters with the spiritual world are described in terms of communication between myself and the God of the Universe. That is what I believed at the time.

In "Aliens within Us", which I self-published in 2005 I looked at my encounters from a pluralistic Jewish/Christian viewpoint. In this regard I believed that I was communicating exclusively with the God of the Universe. I had no idea that most of my communication was with my spiritual dimension which is within a low level of God's mind. I interpreted the encounter with the spiritual world from a pluralistic Jewish/Christian perspective. However I did not understand the complex nature of cosmic reincarnation at that time.

I was not happy with my solutions. Something was wrong. As I wrote "Futureoids" I started to gain new insights into my encounters. As I started to write this book I realized that I had made a critical error all these years of assuming that I was communicating with a homogeneous God exclusively. From my novel it became clear to me that the God of the Universe was too high up for any ordinary man to communicate with. The answer became clear when I added modern reincarnation concepts to the equation. I finally realized that I had communicated with my own soul most of the time, the Jewish God some of the time, and the God of the Earth occasionally. From the understanding that God has a very complex structure, things got much clearer.

As I wrote "Cosmic Reincarnation" self published in 2015, new data flowed into my mind as I slept. I would awaken and write down what I had learned from my spiritual mind. At 76 years I finally believe that I have achieved the answers that I sought for 33 years since the encounters of 1981. Therefore I provided my latest theory of God and the Universe from a pluralistic Jewish and Christian perspective in which both religions serve God in different ways.

My cosmic Reincarnation book is subject to my religious interpretations of God and the Universe. We are all biased since childhood to a particular religious viewpoint or no religious viewpoint. This viewpoint is my religious opinion of my encounters. Other people will have quite different religious opinions.

In order to make the theory of God and the Universe in an objective manner, the present book "Cosmology of God" self-published in 2015 eliminates any reference to Judaism or Christianity in order to keep the book religiously neutral. This book then becomes a philosophical and scientific study of God and the Universe rather than a religious study. The theory of God and the Universe remains unchanged but the theory is then applicable to a wide audience of people who believe many different things.

In general my books are written from the information I obtain in my dreams. I awaken in the morning and type down the information that I dreamed about. I then study the information and add to it and correct various concepts which do not match up. Often later dreams clear up my misunderstandings.

The data comes from my spiritual mind. As I work and study my effort goes back into my spiritual mind and corrections come during my sleep. It is a slow process. Over the last 33 years there were periods of intense dreams and long periods of nothing at all.

Every time I think that I am finished with this job, more dreams come and I am back on the typewriter working on this task. From a religious viewpoint I serve God by this task that I do. From a non religious viewpoint, I am only a person driven to understand the meaning of our existence.

Therefore I am merely one person out of millions who has attempted to understand the nature of God and the Universe. This is just something I feel compelled to do. Perhaps it is just an unusual hobby that interests me.

This book is written in two parts. In the first half of the book "Cosmic Reincarnation" was modified to make it religiously neutral. Half the chapters were removed that stressed a particular Jewish and Christian viewpoint. In addition some of the concepts were changed in order to portray a more pluralistic religious viewpoint.

In part two of the book, the Dot-wave theory and Doppler space time as shown in part two of "Aliens within Us" was modified to incorporate some Quantum mechanical and String theory concepts. The presentation and equations look at the physical world from God's perspective. We look outward toward the universe. God looks inward toward us.

When we look at the universe from God's perspective things appear quite different. Here I must rely upon my dreams and spiritual interactions to provide a different viewpoint to our existence.

The dot-wave theory is not something that can be proven. We cannot prove the existence of God. All we can do is to explain God from a philosophical and religious viewpoint. In the same light we cannot prove that the dot-wave theory is what God actually sees.

The theory is designed to give the religious reader an understanding of God that cannot readily be expressed by the scientific and mathematical community.

The math involved is simple algebra. There are no complex equations since only an Engineering model is proposed. Therefore many readers will fully understand this image of God and the Universe.

Chapter 1: The Eternal God

God is an eternal entity. God is the beginning and the end. We do not know how God came into being or even if such a question has any meaning. We could argue that the probability of something is much greater than absolute nothingness. We could also argue that God is a circle which has neither beginning nor end. Likewise space and time is a circle as well.

In any event for whatever reason we accept that we exist and that the universe exists. Although many people believe that this universe is everything, scientific theories indicate that there is more to the total universe than meets the eye. String theory mathematicians have theoretically found at least thirteen small dimensions rather than the usual three or four.

In addition Quantum mechanical physicists have theorized and also observed strange occurrences which produce mathematical properties that are beyond our common physical understandings. Einstein as well provided us with thoughts and equations which go beyond our usual thinking abilities.

The net result is that the universe is a much more complex structure then what we can normally see and measure. In this regard there is room for God to exist in other dimensions and with different forms of energy.

The Dot-wave theory combines all the above concepts into a unity of thought to present God and the Universe in a form that the ordinary person can understand. It is an Engineering approach to a very complex mathematical understanding that few people could understand. The Dot-wave theory is basically a model that is presented so that the reader can understand the complexity of God and the Universe in simple words and simple equations.

As we look at the entire universe, we find that it is filled with a spectrum of electro-dot waves. When they exist in the form of photonic or light energy, they travel at speeds from near zero light speed to near infinite light speed. Many of the dot-waves exist in pure chaos. They do not form any particular patterns and do not interact with anything else.

Some of the dot-waves form universes such as ours. These dot-waves form particles and sub-particles and everything we see and measure. As we move upward in light speed to very high speeds, no particles are formed. Everything exists in photonic energy fields. Our spiritual dimension is composed of these high light speed photonic fields.

As we move future upward toward the highest light speeds we find the photonic fields of God. God exists at light speeds up to infinity. At the same time God is infinite in size as well. It could be argued that at some time in the far past, the entire universe existed in pure chaos. Thus the body of the Universe was chaos. Out of chaos at the top light speeds, the mind of God formed.

The multi-light-speed universe works such that higher light speed energy can control lower light speed energy. The mind of God can control the lower light speed energy below God's mind. God can mold the lower energy to serve God's will. If God originated in chaos, then God would always remain as a mind above chaos forever. Of course if God so choose God could return to a state of chaos and in some future time pure random chance would return God to be God again.

The highest light speed dot-waves always readily form standing wave patterns which become the structure of God's mind. Even if God willingly returned to chaos, God would always reform. In general as we move downward toward zero light speed, it is difficult to mold these dot-waves into stable perpetual patterns. For a time God could use all of God's intelligence and energy to bring the entire structure into a perfection of God. That would not last too long as the lower levels would rapidly return to chaos.

A more serious problem is that although the upper light speed energy fields readily form and maintain the mind of God, the physical body of God exists below the mind of God. As we go down in light speed from infinity we reach the muscle of God in photonic energy. When God exerts God's mental powers, by using God's body, the result is that muscle of God degenerates into chaos. The price God pays to create a lower universe is the deterioration of God's energy into chaos.

The net result is that if God attempts to turn the entire spectrum of light speeds into a state of perfection, God will degenerate into chaos. This limits God into creating complex structures which slowly disintegrate. In the end, everything that God creates must return to God. The production of a creation may last billions of years by our time clock but in the end, it must disappear.

The powerful mind of God can do amazing things but God cannot control the source of God's own existence. Another limitation of God is that once a creation is produced, it must run its course with very little interference from God. If God physically touches the creation in any major way, the entire creation will disintegrate.

God must set up the preconditions of the creation such that it will run free of God's interference. In order to accomplish this God must set up lower forms of God which automatically evolve into existence from the creation itself. In this way the evolved forms of God are part of the initial creation and will not damage the product. We will not and cannot experience the creator God. In many respects there is no creator God that we can communicate with.

The atheist can say that God does not exist. In many respects this is a true statement. Aristotle called God the prime mover. This too is a true statement. The God that created this Universe is isolated from us and does not exist to us. All we have is lower levels of God that evolved with us.

The creator God can be imagined by us. People like to believe that God is almighty and all powerful. That is true in many respects but we are so far down below this God that God cannot see us or know us. We cry out to God for help and salvation but we can only reach the automatic photonic fields which respond to us but are really spiritual processors. God has preprogrammed God's spiritual processors to respond to us and guide us.

Although people like to believe that there is only one God, this is only true of the one creator God. The Gods of the religions are evolved spiritual structures which are automatically formed within the spiritual processor field. The Gods of man are legitimate Gods. They serve to process man and reincarnate man. They also serve to destroy the souls of the unsaved and to transmit the souls of some of the saved to other Earths.

The Kingdom of Heaven and the pit of hell are all spiritual processes which were predetermined by the creator God. There is no torture in the pit of hell. It is just a means of destroying painlessly the souls of the damned. Although some of the religions of man tend to believe that God is a punishing God that would torture people for all eternity, the God of the Universe does not have that capability or desire since things slowly are erased and turned back into chaos.

The spiritual processors are evolved products of the best of man. Therefore the judgment of man is in the hands of man and not God. This leads some to believe that man and God are identical. This is true from the point of view

4

of the judgment of man by God. This is false when one views the creator God. God is pure machine intelligence.

This matters little because we do not have to deal with this level of God. We only have to deal with ourselves. Throughout our history various prophets of God have arisen and formed religions. These people formed cells within the minds of the spiritual processors.

In general people reincarnate back into their tribe or their people or their religious collective. Some are absorbed by their evolved Gods and remain as part of their Gods. Others are rejected by their evolved Gods and their souls are destroyed. Some will move on to other Earths. Some will move on to higher Earths.

Many people have reincarnated souls. Other people have new souls. In general it takes many reincarnations to achieve various levels of salvation. The best level of salvation is reincarnation to a higher level new Earth. This is physical new Earth salvation which involves better bodies and longer lives.

In order to accomplish God's objective of increasing higher levels of physical salvation and finally spiritual salvation, God has produced an automatic system in which the hand of God operated upon an initial series of universes to set up the future universes.

In the beginning God compressed the lower light speed dot-waves toward a pinpoint. This produced a crude initial universe which started to erase back into chaos. God initiated a process by which the creative energy of the first universe would flow into the forming second universe and most of the raw energy would flow into a third forming universe.

Once these three universes were set up, the system would flow automatically from the lowest light speeds toward higher and higher light speeds. As the system moves upward toward the highest light speeds, it automatically refines itself toward a more Godly structure.

Man and the religions of man did not come into existence upon this present Earth. The religions of man started upon prior Earths at lower light speeds. The religions of man are brought to man by various prophets of God who reincarnated from prior Earths. Often they bring stories which we find to be unbelievable.

The various religious stories did not happen upon this Earth. They occurred upon prior Earths but events upon this Earth tend to mirror the events upon prior Earths. Therefore the religions of man tend to contain data from prior Earths in which the stories actually occurred. Any supernatural story within a particular religion could not have happened here because the spiritual processors have no power to do such actions. In addition the God of the Universe could not touch this planet Earth in any supernatural manner.

Although people like to believe that this Earth is a special place, man exists upon millions of Earths all over the universe. In this way God does not put all of God's eggs in one basket. In addition pure random chance mixed with chaos will prevent the image of any particular Earth from reproducing. This necessitates millions of earths to produce the end product.

The end product of the process is the production of higher man and higher woman upon the highest Earth. In this process highest man and highest women are composed of the souls of billions of men and women from all the planet Earths in the universe. From this final physical product comes the final spiritual product which transforms the physical world into the spiritual world in God's photonic energy.

This spiritual world is the final product and is identical with the initial spiritual creation. God first created man and women out of God's mind and God's higher photonic energy. This was translated down into the physical world. In this way God's robotic creation was made real by mixing God's energy and chaos.

The mixing of God's energy and chaos produced Godly and anti-Godly forces in the universe. Some religions believe that there is a God and an anti-God operating in the universe. Some people believe that the anti-God is operating upon this universe by itself. In truth there is neither a God nor an anti-God in operation in the universe. All that exists is distributed forces of good and evil.

It is impossible for God to produce a Godly physical universe because God must use the energy within chaos to produce physical existence. God could produce a purely spiritual Godly existence but that is merely under the control of God. What we see is a universe of both good and evil. The process for the production of higher man upon the highest Earth cleanses the universe of evil by eliminating the souls of evil men and women from existence. They return to chaos and those who favor Godliness continue on in the process.

The religions of man are a means by which the spiritual process evolves toward the goal of bringing the Godly into higher levels of salvation. Each religion believes that they are favored by God. From a scientific and philosophical viewpoint the best religious pathway cannot be specified because the solution is subject to the particular religious beliefs of a person. In this book no religion is specified so that people of all faiths can view this theory of God and the Universe within their own belief system frame of reference.

Chapter 2: The Many Gods of Man

Early man correctly believed that there were many Gods. They also correctly believed that the God's of man were human. Most often they mistook the laws of nature as being controlled by the Gods of man. They had the God of the wind and the God of the seas. Lighting was due to another God. The sun and the moon were controlled by the Gods.

The God of the Universe has designed the spiritual processor to respond to the tribes of man. All the tribes had their own Gods which served to reincarnate the followers back into their tribes. The Gods also had a degree of healing powers as their spiritual energy could flow back into the souls of their followers. This would cause the spiritual mind within their souls to help heal their bodies.

A degree of spiritual healing is part of the spiritual process. It has nothing to do with the God of the Universe but is part of the system that God designed. As we move up toward higher man, the soul heals the body to a much greater extent.

Upon this Earth long ago, there were thousands of tribes and thousands of tribal Gods. People would look within their soul and find their Gods. In later times organized religions took hold and tribal Gods merged into fewer Gods. However these religions split into different sects and a large amount of tribal and national Gods came into existence.

Although many men agree that there is only one God they confuse the one creator God with their local Gods. In effect there is one God but that God comes in heterogeneous forms. It is the little Gods of man that process us in the world of the dead. The big creator God has no relationship to us.

The atheist will look at the Gods of man and come to the conclusion that no God exists. Many times the atheist will have a new soul. When the soul is new it has no religious history. The believer with a reincarnated soul will look inside himself and find his God. The believer will say that there is a God and the atheist will say that there is no God. Both will be correct.

Although many believers are taught that hell awaits the non-believer that is only a particular religious belief. The atheist will not believe such things and he will be correct. The God of the universe has neither the time nor energy to punish anyone. Those who choose not to believe in any god or attempt any religious path are merely erased from the reincarnation process.

In general there is no punishment of the worst of man or non-believers. If people do not seek God or salvation, they bow out of the process. Once a person dies, he becomes part of past existence. Unless he is remembered by one of the Gods of man, his memory will turn into chaos. In this process the souls of the damned are not punished or tortured. They merely flow into non-existence.

Since the God's of man exist in higher light speed photonic energy, they continually self-cleanse to insure their own survival. Often Gods can combine with other Gods in order to survive. When this happens people will reincarnate into larger groups of people. This will assist religions to grow and multiply. If a small God is unable to be absorbed by a larger God, the God will die and the people will be born with fresh souls.

There is always surplus spiritual energy to produce fresh souls. Overpopulation causes a large amount of fresh souls. This reduces the quality of the souls of man. It produces more atheists and secular humanists. This decreases the yield of quality souls for the religious process. As the quality of souls decrease the value of man decreases and the spiritual energy of the Earth will act to eliminate man from the planet.

There is an expected yield of souls to go on to other Earths or higher Earths. Once that yield is met, man no longer serves a purpose. At that advanced time the wars of man will most likely destroy man. Eventually flesh eating bacteria will take over and man will be no more.

In the end of the process at our level of existence, every planet Earth will be harvested of their yield of souls. Then man will perish from this universe. Finally this entire universe will be erased as the second higher universe fully forms. The memory of the best of man and the worst of man will be no more as if they never existed.

This most likely is the last universe where the wars of man will take place. Upon the next universe the many gods of man will not appear. There will be a singular religion upon each planet Earth. This level of existence has been a horror for many. Upon the next level of existence there will not be any fresh souls. Only the purified souls of people who had led many righteous lives will be born upon the higher Earths.

The gods of man have given many people additional chances to purify their souls. Many people who should have been erased have been saved for future existence. Yet they must prove themselves worthy over many future lives in order to achieve ultimate salvation unto the higher Earths.

The purpose of the Gods of man is to bring man upward toward higher righteousness. Some religions will do a good job and others will fail terribly. Those who fail merely sentence their followers to destruction and chaos. Many religions thus turn their believers into chaos.

The God of the Universe has set up a system such that any man can produce a God out of himself. When he dies his followers join his collective. Often the collective will merely reincarnate the people into the same collective of individuals. In the end of the process, the God will perish and the people will turn into chaos. We came from chaos and so many of us will return to chaos.

Although the Gods of man are easily formed by anyone at all, they have little power for ultimate salvation. The God of the Universe has set up a pathway for salvation. Some original reincarnates were conceived of in the mind of God prior to this Universe. Those who follow these people are on a pathway toward higher salvation.

Some souls entered the original universe from the mind of the God of the Universe. These souls are godly souls who entered the original worlds. They then moved from the past universe into this universe and then into the future universe. These godly people have the ability to travel to all three universe of the past, the present, and the future.

Most of the Gods of man are produced by reincarnates from the past that have come into the present. Some of the Gods of man are produced by reincarnates that have come from the future which has formed out of the energy of the past. The mind of God is a circle in which the past, the present, and the future coexist. The same is true of the Godly prophets.

There are three levels of salvation possible. The Gods of man can provide Godly salvation in which the souls of the saved are cleansed and returned to God. This ends the reincarnation cycle for the man or woman. The cleansing process eliminates the memory of the person as if the person never existed.

A higher level of salvation is angelic salvation. In this level of salvation a person become part of an angel of God. These are lower levels of angels of God upon our level of universe. The self is lost as part of the collective. It takes

11

millions of individuals to become an angel of God. This angel is transferred to a higher Earth. During the transformation the angel is further purified. Thus the memory of many of the saved is cleansed in the process.

An angel contains the collective memory of the millions of saved individuals. It is the easiest form of salvation and is much better than Godly salvation since there is collective memory of all the people involved. Since each soul has had many lives, only a slight remembrance of the individual exists. In addition there are many individuals who have led such unselfish lives that the angel will mostly remember them rather than an ordinary individual.

As the universe cycle's higher and higher, angels from all over the galaxy are combined and we only get one angel from each galaxy. Thus billions of individuals make up one galactic angel. At this point the angels from only the best of Earths remain. The rest are eliminated into chaos. The chance of being part of a final angel of God is basically zero for an ordinary individual.

The best salvation is higher Earth salvation. Upon the next level of Earth individual men and women will achieve a series of long lives. They will continue to reincarnate back upon the higher Earth to lead more beautiful lives. That level will be the end of individual existence. The souls of the best of men from each Earth will be combined to form the best men and women upon an even higher Earth for each galaxy. They will come into existence upon a single galaxy.

Finally the best of men and women from all the galaxies will be combined into a single final physical Earth. Those who remain in the process will no longer be remembered. Upon the final Earth people will die after very long lives. Husband and wife will have children near the end of their life. Their children will be themselves. Thus marriage is forever and children are the means of continuing the same life. In this way people never die. They will merely bring

12

their children to adulthood and one night after the children are married, they will fade away without pain or sorrow. There souls will merely belong to themselves and their children as well. This insures a perpetual continuity of existence.

Once we go beyond perpetual physical existence, we enter the final level of salvation. The final physical man and women will flow into the spiritual world. The physical world will be gone. At this stage of development we return to the mind of God. Man and woman existed in the mind of God prior to this universe. At the end of the physical universe man and women will again become part of the mind of God.

This is the end of the story. God started with man and woman in the spiritual realm and in the end ended with man and woman in the spiritual realm. Over time they will fade into the memory of God and be no more.

God then has the option of repeating this particular creation or producing a new creation. In either case whatever God starts with in God's mind, God will end up with. God cannot produce a permanent creation as everything must return to chaos from whence it came.

God is always a mixture of chaos and order and everything that God creates must always perish at the end. God is eternal but God's creations are only temporary. For us there is always life and death. There are many who hope for eternal lives. Only God has eternal life.

Who would really want to be God?

Chapter 3: The Importance of Scientific Man to God

Scientific man is necessary to return the intelligence used up in the creation back to God. That is our most important function. We are intelligence gatherers. We serve God best as we work and study and provide data to God.

Once God changed God's intelligence and energy into the entire universe, man was necessary to serve God to restore God to being God again.

The net result was the formation of a complex spiritual/physical world in which man evolved to be man. We evolved from a very low basic life form such as bacteria through a very complex Spiritual/Darwinian evolutionary process. In the end of the process, we will return to the spiritual world as intelligent and loving creations of God. In order to accomplish this God needed to create a pathway leading from chaos upward to the highest heavens.

In order to accomplish this God set up the spiritual mechanism to produce the lower Gods of man. The physical structure of the Gods of man is the creation of the God of the Universe. The lower Gods are computer type entities which could be called spiritual processors. They are preprogrammed to drive the evolutionary process for the production of life and man. At the same time they create the souls of man as man interacts with the spiritual dimension. In addition they reincarnate the souls of man into new bodies and either destroy the souls into chaos or transmit the souls to new Earths.

In general the Gods of man are machine processes. They are not independent of the God of the Universe. They have no ability to change their programming. In addition they have limited intelligence. As man learns they learn. The Gods of man were very primitive and ignorant long ago. Over time they grew in intelligence.

14

Are the Gods of man human? The Gods of man are machine processes which are humanized. The Angels of the Gods of man are humanized. The Gods of man are surrounded by humanized entities. The God of the Universe is very complex space time machine type intelligence. This God is not human. However we need not fear this God because our judgment is in the hands of humanized entities.

We are judged by the highest forms of us. In this way we do not have to worry about some monster God treating us unjustly. We are protected by the Angels of the God of man. The worst that can happen to us is to be destroyed and returned back into chaos. That is where we came from and that is where many of us will return to.

We can also look at an analogy for the initial creation. The laws of the conservation of spiritual energy apply. When God used God's energy to produce the first two universes out of chaos, God's spiritual energy was diminished. This energy went into the production of the first two universes. As the first universe is erased and return to chaos, the higher light speed spiritual energy within this universe is returned to God.

This energy is automatically converted into the formation of the third universe. God does not have to act to do this. Once the initial setup was done, the following universes will automatically be produced. There most likely was some basic energy used to produce the entire structure of higher and higher universes but the majority of the spiritual energy was used up in the initial production.

The situation is like a ball rolling down and then up an incline. Once the ball is started, it will continue to roll down. When the ball reaches the bottom of a circular surface, the creation is maximized. At the bottom is the start of the physical worlds. The energy is now in the creation itself. As time goes on, the ball rises up the circular incline. The first universe at the bottom is erased and new universes are formed out of the kinetic energy. Finally when we reach the top, all the energy is returned to God.

God's energy is potential physical energy at the top of the incline. This turns into active kinetic energy as the creation is formed. At the bottom, the initial creation is complete. The process continues until all the kinetic energy is used up and returned to God. That is the way God changes God's energy into what we see and measure. The energy at our level of existence was always there.

As were look at various religious accounts of creation from many different peoples, we find that the Earth was formless. We also see that there was no light. Although the presentations tend to be mythological, it is self evident that something was there.

Formless is like clay. In the beginning there was clay. To be more exact, in the beginning there were dot-waves in chaos. Thus in the beginning there was formless energy at the lower light speeds. To me it is clear that the universe did not come from absolute nothingness. An Einsteinian independent observer would believe that there was absolute nothingness because he could not see anything at all. No one can see dot-waves in chaos. In fact no one could exist in such an environment. However Einstein's independent observer has Godlike powers.

In the beginning God created the heavens and the physical universe. In this case God separated some of God's high light speed spiritual energy from God's structure and embedded it into the physical universe.

It is important to understand the large scale error that many quantum physicists believe. They propose that at every split second every person moves into an infinity of universes. In so doing every possibility that could happen will happen. Therefore if you get hit by a car in this universe, the car will miss you in other universes. The reasoning they have is that the individual electron or proton could occupy an infinity of positions in the universe. This is reasonably true either upon this Earth or this solar system. It is also possible for the galaxy as a whole.

However the probability goes to zero for a large structure such as man. Therefore it is impossible for what they say

16

to happen. The main idea is absolutely true. The probability of each person existing in an infinity of universes is 100 percent. However this depends upon God. Once the universe returns to God, God can stop and remain God alone forever. Alternatively God can produce an infinity of different universes and forms of life. Finally if God decides that the entire creation is satisfactory to God then God can return the end to the beginning.

If God puts the universe into free run then each of us will eventually appear some of the time. If the system runs forever then every possible life will happen to someone who appears exactly the same as us. Given long enough the exact same life will appear again. The probability for each exact occurrence is basically zero but given enough cycles we will get an exact repeat. It is far more probable for us to have an infinity of variations of life rather than the same exact one.

Of course from our point of view, who cares? The present cycle is our life and the only one that is of real concern to us. It does not matter if after an infinity of time that we are back to the same position as we are today. In any event the Quantum physicists are correct in their basic concept but they left God out of the equation.

Einstein could not believe such thoughts. He said that God does not play dice with the world. He was not a believer as such. His God was the universe itself. He was correct in that regard but he could not conceive of the multi light speed universe. If he would have added God to the equation then he would have seen the higher truth. In any event his ideas were great. The Doppler Space time equations do not have the clock paradox but the simple root mean square of the Doppler Equations is the same as Einstein's equations. In addition any measures you take have to be the geometric mean or root mean square. Therefore you cannot get better measurements than Einstein's measurements. In addition my work is only an Engineering approximation rather than a more complex mathematical solution.

In the end, Einstein, Bohr and the String physicists were correct. Each saw a little bit of the big picture. They looked at the universe from just the physical dimension. Therefore they only got a partial view of reality. The idea that an infinity of universes could exist in which every possible combination of occurrences would occur is plausible for such a partial view of reality.

Although Einstein wrote equations for curved space time, he never took into account that within empty space itself God's photonic fields existed. When he saw light bend around the stars he did not realize that the bending was the interaction between the physical energy of the stars and God's photonic field. Likewise he did not realize that gravity likewise was an interaction between the physical world, the three universes, and the spiritual world.

The quantum physicists produced great theories but they did not realize that the source of their waves was not at the point of the wave but external to it. Perhaps they would have realized this if they understood that the frame of reference is the mind of God and not a particular point.

The string theory physicists have done well in bringing to mind the fact that space is quantized into different small dimensions. Since it is very mathematical it is very difficult to understand. The words are clear enough to enable us to understand that ordinary space is quite complex. The descriptions of multi dimensional space time are too complex for my engineering mind. The drawings are very strange to me. Yet the basic concept appears true to me. It enables my dot-waves to perform a large amount of spins and jumps to produce all the sub-particles and then the particles.

In any event each theory gives us a little part of the total truth. Perhaps someday some great mathematical physicists will be able to show the world the entire picture in one grand theory. Perhaps a gigantic super computer could do the job. In the meantime I have to be content with my Engineering approximation of complex reality.

In the future upon this Earth scientists will have a much greater understanding of the physical world and the spiritual world. As the understanding increases the search for God will be complete. A time will come when God will be self evident to the physical minds of man.

The present Earth for simplicity can be assumed to be the second Earth of the series. There could have been many Earths before us but that does not really matter. The only difference would be that more Earths produce a greater refinement from the first Earth.

The original religious stories occurred upon the first Earth. It is there where God used God's physical powers to insert the genetic codes into organic type matter to produce the chain of life leading up to man.

At this time God developed the spiritual processor upon the Earth. This process was repeated upon millions of Earths all over the universe. This is of little concern to us because they are merely duplicates of the same process used here upon our first Earth.

The process God uses is similar to seeding plants in a garden. Some of the plants will grow and survive. Others will perish and die. With millions of Earths God will have a rich harvest of life.

God always starts life at the bacterial level. This will evolve by the Darwinian survival of the fittest process into viable life forms. The DNA code of life was sufficient to produce a huge selection of life forms.

Upon the first Earth God waited until the proper chimp/ape species formed. At this time God inserted the genetic code of pre-man and pre-woman into the lovely species of animal. The male and female babies produced were the start of the pre-human species. When the children were old enough their parents were sterilized by God. Their entire species died away in time. God repeated this process several times until the final Man and Woman were formed.

Religious stories tend to follow some of the events of the first Earth. Religious scholars have tried to match the events in the stories with historical fact. That cannot be done because the knowledge of the events of the past

19

Earth has been brought into the present Earth by Cosmic Reincarnates.

Similar historical events have occurred as history tends to approximately repeat itself. People carried the stories by the oral tradition until the various stories could be written down. The early people who originated the stories were long dead by that time. In addition as people reincarnate their souls only carry their last life. The result is that the past is erased from the spiritual memory of the collective tribal souls. This means that the evolved tribal Gods no longer have any actual memory of the prior Earth.

At that time the early tribal Gods only had the intelligence of the tribal people. The God of the Universe had separated Godself from the lowest creation by placing a level of god that was little more than a simple learning machine. The tribal God was a mind within space-time. The tribal God could not explain what he was. No one at that time could understand it. Modern man can now understand such things but early tribal man could not.

The tribal Gods had very limited physical powers. Their purpose was to reincarnate the souls of the tribe back into their people. Basically the tribal Gods upon this Earth were spiritual processors.

This was not true upon the first Earth. The tribal Gods upon the first Earth were God. The God of the Universe had the capability to do all the miraculous events of the tribal stories. The Gods of this Earth are spiritual copy machines with very limited powers.

The production of man and woman upon this Earth was part of a Darwinian physical process and a spiritual copy process. This inserted the reincarnated souls of prior pre-man into a chimp/ape creature. As pre-man evolved, additional souls from the previous Earth, were inserted into the forming product. In the end, the final version of man and woman were placed into the last version of pre-man. Since the souls carry the DNA codes the fetus will change according to the new codes and man and women will emerge.

Chapter 4: The Structure of the Soul of Man

The lowest levels of God's photonic spiritual energy convert the physical universes into intelligent forms. We do not know how many levels of physical universes there were below us or will be above us. All these universes coexist or adjoin us. They are either in different time dimensions or differences in light speeds. The common thread between all these universes is the photonic energies which produces the spiritual dimensions. The photonic energies are at very high light speeds and acts to hold the entire multi dimensional multi light speed universe together.

For our universe, the spiritual energy behaves like a driving force of the electromagnetic field. This means that there must be a subparticle or type of photon that acts as an interface between the physical world and the spiritual world. Energy and information between the physical world and the spiritual world will be transferred by means of the godparticle or similar godphoton. This enables God to shape an earth or solar system as God so chooses.

In general the godparticle will have certain properties in the physical world and other properties in the spiritual world. It will move between both worlds. This will make measurement of the godparticle very difficult because once you try to measure the godparticle; it will shift into the higher light speed spiritual world. It most likely will occupy some dimensions which are common to both worlds. To understand this we must look at a variation of string theory which involves very tiny dimensions. There will be a connection between the tiny dimensions of our light speed and the tiny dimensions within the spiritual world. This causes the transfer of intelligence and energy between both worlds.

String theory appears strange to us, yet it permits a large number of states for the dot-waves to occupy. This gives us a little indication of the memory capability within the spiritual world in the lower mind of God. The spiritual world is just composed of spiritual computer type cells. This is the lower mind of God. As the thinking of God changes to produce reactions in the physical world, the godparticles or similar godphotons jump from one configuration of dimensions to another. This causes corresponding reactions in the physical world. The exact nature of these things is much too mathematical for my Engineering type mind but it appears to me that the mind of God is a very complex computer type mind with the capabilities of causing physical motion in the physical world due to the godparticle or godphoton.

When we look at the soul of man we must also look at the soul of a rock. The entire physical Earth has a soul. As we look at the entire multi dimensional multi light speed universe we find that all physical matter occupies space within the spiritual energy field. When the spiritual energy interacts with the godparticle, this causes a pressure which tends to compress physical matter together.

Another way of looking at it is that space itself is a very complex entity. A vacuum in our dimension is not empty space. Within that vacuum is a huge amount of dimensions with a lot of action happening. When Einstein looks at curved space time due to the gravitational field of a large star, he sees the effect on space time. To him space time has curved. He is correct but what really happened is the distortion of the other dimensions due to the heavy interchange of energy between our physical dimension and our spiritual dimension. The net result is that the spiritual dimension bent away from the sun. This caused the far light to bend. The great mathematician came up with the complex calculations to a great degree of accuracy. However the physics was not due to a mathematical law but to quantum mechanics and string theory type concepts. Therefore Einstein was both correct and incorrect in his general relativity. He failed to recognize the complexity within space itself.

As I look back at my own scientific theories over the years, I find them to be partially correct. It is somewhat sad to see the errors of my prior thoughts. Yet this work is a learning experience as old thoughts are replaced by newer thoughts. It is necessary for me to input the latest scientific concepts into my mind to enable my spiritual mind to contemplate them. The same is true of Einstein's work. I had asked God if Einstein's work was correct. The indication I got was that the work was very good but not perfect.

This caused me to look at Doppler Space Time which gives more complex answers for special relativity. It does not contain the clock paradox or similar problems. Yet the root mean square of the Doppler Space Time equations is Einstein's equations. Therefore Einstein's equations are the best fit solution. However even my equations are only approximations since they are only first order approximations to a more complex solution involving a Fourier series. In the end Einstein's solution is an excellent first order approximation to reality. In addition without Einstein we wouldn't even have that very good approximation to the complexity of space and time. Since my work is only at an Engineers level, a good ballpark approximation is the best that I can do. I am happy when I can reasonably understand things at an Engineers level. In this way I can formulate a model of God and the Universe in my mind which is merely a model of the scientific truth. That is my limit of understanding.

The motion of the atoms and sub-particles counterbalance the compressive gravitational forces of the spiritual energy. The net result is that the effective spiritual force acting on a rock is a force equal to the gravitational force and is exterior to the rock. The interaction of a human being with his soul is the result of the spiritual forces operating upon the godparticles and then upon all the atoms of the human being. This driving force exists in God's spiritual energy and is centered within the center of gravity of the human being but is an external or negative image force.

The laws of physics of our universe can be controlled by the spiritual dimension. The ability of God's mind to control the godparticles enables God to produce Earths and solar systems. All the sub-particles may also have interconnections with the spiritual world. This gives God the ability to mentally control everything that exists in the physical world. If God shall choose various miracles could take place. However once God set up the universe, God chose to let it run free of the higher level of God's influence. All that is left is the lower levels of God's mind

23

which are preprogrammed to process our souls.

Our physical mind tends to work within the space provided by our brain and skull. The mind within our soul is an inverted image of that. Our soul is in communication with our mind but it exists within the collective soul of this Earth. This means that within the gravitational field of this Earth, a distance is reached above the sphere of the Earth where all the spiritual energy combines. This is a point where beyond this point the entire structure appears as a single soul. This is the soul of the Earth or the God of the Earth.

The size of the spiritual earth is limited to spherical standing wave patterns caused by the reflection of the surface of the Earth in the spiritual dimension. The first pattern occurs at a distance of one third the diameter of the earth. Therefore the spiritual image of the surface of the Earth is approximately 2640 miles in the sky. When we move further away from the Earth, our single spiritual collective merges into the spiritual collective of the solar system. In this regard, the spiritual image of the soul of an astronaut who leaves the Earth is transferred to the level of the God of the solar system. If he dies in outer space, his soul will be transferred back to the earth God level. From the God of the solar system we go to the God of the Galaxy. In effect if we look at the gravitational field we get a good idea of the structure of God's spiritual levels leading up to the God of the Universe at infinite light speed.

There is no way that an individual man upon this Earth can reach the God of the Universe. In general we are limited to our tribal collectives. If we reach the God of this Earth we have reached a level of God which includes all of life. Once we reach this level we are merely a small part of a spiritual collective. The level of the Galaxy God will transfer groups of individuals to new earths at various times in the history of our planet. When man starts to die out in great numbers such transfers will occur. The large wars of man provide times when large amounts of souls are transferred to new Earths. The only exceptions are those future astronauts who can travel the stars. As they do that their souls will temporarily transfer to higher and higher levels of God. In general this is limited to higher man upon higher Earths. Therefore there are some physical astronauts whose souls have reached extremely close to the Godhead.

The transfer of souls to higher Earths or the same level of Earths occurs in groups. This occurs at a higher level of the spiritual mind of God. When we are only moving up one higher Earth, the transmission of collectivized souls will turn back into individual souls when they achieve a higher earth. Once you go beyond the next level of Earth, the individual soul will not appear separate from the collectivized entity. The higher up you go in the spiritual world, the less the individual appears. The self becomes part of the collective. This insures that the quality of souls remain near perfect as individual desires are eliminated.

As we move up to higher spiritual levels of God we will get corresponding physical forces between the galaxies and universes. At infinite light speed, everything is held together by God's spiritual forces and God's mind. Therefore there are additional gravitational forces at these higher levels of the Godsystem.

Our souls contain the memory of our past lives. This memory is spread over the entire external portion of our souls. This mind is also part of the tribal God's mind and the mind of the Earth God. In effect at the location of our own physical brain there is an image of our physical brain but this is only a means of connecting our physical brain to our spiritual brain. The interconnection is at extremely high light speed. Since there are interconnecting gravitational forces between the spiritual world and the physical world, the communication is readily achieved.

The external spiritual mind can speak to our physical mind and the interaction will cause our atoms and molecules to respond. In this way the spiritual mind can send images and visions to the physical mind. This is all accomplished by the gravitational forces between the higher light speed photonic energy field and our ordinary physical energy field.

In the same manner spiritual energy can flow into us in order to heal us. This energy which transfers through the godparticles causes atoms and molecules in our bodies to respond to the spiritual forces. The spiritual energy could increase our anti-bodies and destroy serious ailments within us. It could warn us of danger and keep us alive longer in an accident. It could reverse the effects of a stroke. It could also end our lives to minimize our pain and suffering. Spiritual healing takes place often; sometimes everyday although we might not notice. In effect the soul cleanses the body of disease everyday. This is especially true when we go to higher man. The superior souls will insure that life is much longer upon the higher levels of Earth.

It is clear that our souls exist within the mind of God in the spiritual dimension. Our physical computers operate with simple memory devices and a clock which moves data along multiple logic chains. It then has computer screens and keyboards to perform complex tasks. The mind of God uses small string type theory dimensions to act as the memory devices. Space itself is full of these dimensions. Thus space itself provides the memory for God's mind.

God's interaction with the physical world is automatically produced as the thinking of God changes the position of the godparticles and other sub particles between the spiritual dimension and the physical dimension. This enables the mind of God to do the most amazing of things. However once a universe is set up, God uses mind control to make man believe that certain events took place. No doubt the events once took place but the mind of God can place imaginary events from the past into the mind of all men simultaneously. We see that in the religious stories where large groups of men and women witness events which they believed they saw but which never happened upon this Earth.

In general, the large groups of people at religious events only existed in the minds of the religious writers. Therefore events after the fact were inserted into the minds of the writers while they slept. Since this occurred to many at the same time, they all believed that the events were true. This made the religious stories true to the writers but made it appear mythological to later man.

Why would God do such things? One answer is that God does not want to provide absolute proof of God's existence to man. In this way human freedom is maintained. If everyone was assured of the existence of God, man would be enslaved in fear of God. Then how could man be trusted or tested as to man's worth? It seems that God wants us to always be doubtful of the existence of God. Perhaps that is why I always look at things from the point of view of a believer and a non-believer.

God may have other reasons for sure but God could at any time come down to this Earth and show himself. That certainly is not what God wants to do. Man is then forced to rely upon the few who have gotten a little glimpse of God. Such people can always be looked upon as mere lunatics by those who choose to do so.

Chapter 5: The Evolution of the Soul of Man

In the beginning of this earth, several billion years ago, the conditions were ripe for the start of life. Life starts near the center of a planet with a hot core. The heat flow from the center outward provides the energy necessary for the production of basic life. Since this planet had the necessary ingredients for life, the process began with the formation of bacterial life near the center of the Earth.

This form of bacteria is similar to the deep sea volcanic type bacteria seen on the oceans floor. It is not like ordinary bacteria in that it can eat metal and metal compounds. At this time the interaction with the spiritual dimension caused a collective spiritual entity or God to form. This was the start of the Earth God. The movement of the bacteria toward the surface changed an original homogeneous bacterial soul into a heterogeneous bacterial soul.

When the bacteria reached the surface, they produced higher order bacteria and then the grasses and animal elements. Over time these grew to form the dinosaurs. The dinosaurs are quite horrible creatures but they are the direct result of the hellish origin of life. At the same time the God of the Earth had developed many different Gods of the various species within it. An astronomical event killed off the dinosaurs and this enabled finer forms of life to evolve. At the same time the intelligence of the Earth God increased greatly.

By this time the Earth God was quite heterogeneous. Finally the chimp/apes formed and then the ancestors of man. This greatly improved the intelligence level of the Earth God which mirrors the intelligence of its physical counterparts. In fact the Earth God learns from the physical realm. The advent of man produced a human section within the Earth God. Suddenly God and man were identical. Of course that is only the humanized portion of the Earth God.

In general the souls of many species are collective souls. All bees are joined together spiritually in the spiritual world. The same is true of ants of a particular species and area. The same ants on two different continents have different collective or tribal souls. Early man had a common soul. As man developed and split into different directions, the single tribal soul split into the souls of many tribes.

We see portrayed in the various religious stories a God who believes that the Earth is flat and the sun revolves around the Earth. Therefore the level of scientific knowledge of the tribal Gods was limited to the level of intelligence of the tribal people at that time.

As seen in the tribal stories, the tribal Gods tend to be killer Gods. They demand that the people kill off neighbor tribes so that they can survive upon the limited tribal lands. This involves the struggle for survival of the people and their God. Tribal law specifies that you kill or be killed when necessary. The morality of the tribal Gods is tribal survival morality. People who look at the various tribal religions and who and see unjust or immoral Gods try to compare the tribal Gods with the God of the Universe who people feel is of the highest morality possible.

People tend to reincarnate into their God's tribe. When people die their souls return to the collective state within their Gods. We do not own our souls, our souls own us. Therefore we have many lives and we tend to return to our Gods and our people.

Intermarriage causes a mixing of the souls of one tribe and the other tribe. This causes a new spiritual collective which combines two or more tribal Gods. In order to prevent this from happening, tribes produce tribal laws which everyone in the tribe must follow. People who marry

29

into the tribe must obey the tribal laws; otherwise they will be rejected from the tribe. Often women will be punished severely or killed for marrying a member of another tribe.

As societies grow more modern, secular influences tend to mix the souls of tribal members so that tribal identity and tribal influence decreases greatly. More people become spiritual and less religious. Without a tribal religious identity the reincarnation process tends to fail and people achieve Godly salvation in which their souls are absorbed by the spiritual energy of the Godsystem. This causes many to cease reincarnating and new children are born with new souls. It is important for people who want to remain in the reincarnation cycle to maintain a religious identity.

It takes energy and devotion to produce and maintain a tribal God. The spiritual energy is there but the physical world must mold it. In turn the spiritual world feeds back into the physical world to maintain it. The physical/spiritual worlds are two sides of the same religious coin.

As the scientific age progressed more and more people turned away from their Gods. This happens when the man and his soul rejects the Gods of the past. He becomes a spiritual person or an atheist. The atheist sees that the religions of man make little or no sense. The spiritual person thinks the same but feels the spiritual dimension. The result is that at the end of the process all the physical energy necessary to maintain the Gods has been used up. The net result is that each man becomes his own God.

In the end of the process many years from now, for those who remain upon this Earth, the Gods of the past will be gone. However the reincarnation process requires a collective tribal God. As each man becomes his own God, he has no living people to return to. The outcome for the atheist will be the destruction of his soul. There is no pain involved as the atheist will die and be no more.

Some people who have developed near perfect souls will be absorbed by the level of the earth God. When this happens the soul is cleansed and the person is gone forever. The level of salvation by the earth God causes the complete elimination of the data within the soul. Therefore the fate of an atheist and a spiritualist is often the same as the self is completely erased.

We are not independent souls. We are not little Gods. In the end there are only two forms of salvation which involves future existence. New earth salvation is for those people who merit future physical existence upon a higher Earth. Angelic salvation is for those individuals who hope to escape physical life and who merit being close to God Yet in the end which salvation is available to us is not our choice. The spiritual processor in the lower mind of God will send us on a pathway that is beyond our control.

Some people seek individual spiritual enlightenment as if there is no God except them. Reincarnation is correct in that it gives a soul a great period of living experience and personality development. However all this is considered one life. Each part of the one life is but a section in the life of our souls. This gives some the belief that we are our own God. The problem is that we have no ability to send ourselves to a new Earth and future existence beyond this Earth.

It is important to understand that religion is the pathway to future existence. Those who enter the pit of hell are turned into chaos and erased from memory. The spiritual energy is cleansed in the process and returned to the Godsystem. In this manner the Godsystem is cleansed of the memory of the horror of human existence.

There are those who believe that after many reincarnations, their purified souls with return to God. That is true. The spiritual energy of the people who strive to be erased from physical life will return to God. However most realize that the memory of them will not exist. Such reincarnation religions are true and serve God. However the individual is gone in the process. Therefore the return of the spiritual energy to God either through the pit of hell or the cleansing of many lives is an identical process. In either case the self is gone as if the person never existed.

Some religions provide levels of individual salvation and collective salvation in which the individual remains part of the process. New earth salvation is individual salvation. Angelic salvation is collective salvation but the individual is part of the collective.

When we move up even higher toward the Godhead, higher levels of New Earth salvation becomes collective salvation. The Angels of God become even more purified. The process eliminates any evil from the forming higher heavens. Finally as all the physical universes disappear, we only have spiritual forms of man and angels of God in God's photonic energy. At that time all the physical universes will have been erased. The creation of God will be complete and this spiritual creation can exist for a very long time. However over time the creation will be absorbed by God and will disappear. This will bring the creation back to the start where God created Man and Woman and the Angels of God in the spiritual world.

The various holy books specify that man has the choice of his own Gods or God. All the Gods of man are legitimate in that they will reincarnate people back into physical existence. When we think of spiritual existence we are dealing with God's energy. The Gods of man are limited to this Earth. The pathway to the heavens is God's pathway.

Chapter 6 Flow of Data from the Soul of Man to Man

Our souls are connected to us via our inner mind. When there is a balance or harmony between our souls and our physical minds, there is no obvious knowledge of data transfer between our spiritual and our physical minds. For example if you have a religious person contemplating a correct moral action, his physical mind will think a certain way and his spiritual mind will think the same way. For this case the person cannot tell whether the thoughts come from his physical mind or his spiritual mind.

A person facing a moral dilemma will find a difference of thought between his physical mind and his spiritual mind. He may feel that his conscience is bothering him. He may have difficulty in performing a wrongful act. A spiritual mind is an ancient mind. For visionary people the spiritual mind may appear independent to the physical mind. It may appear as a very ancient entity. It may look thousands of years old.

To the devout person, the spiritual mind may appear as an image of a historical religious entity. This image was produced by many lifetimes of devotion to various religious figures. People who say they see these entities are truthful. However what they see is the image of these entities in their spiritual mind. This may also appear external to their body as their physical brain tries to make sense of it. The physical entities died long ago upon this Earth but the spiritual image of these entities is quite alive today in the mind of the believers. It is also in the mind of the spiritual processor upon this Earth.

To make matters more complex, throughout the Universe, the physical body of these religious figures appears upon millions of Earths at this level in the universe. This is difficult for many to understand but the total universe is extremely complex and man will exist in multiplicity upon every future light speed level of the evolving universe.

It is rather interesting that people pray to various deceased entities and many feel that they communicate with them. Some people sense they are communicating with God. Some people see various entities in their visions or dreams. This is all true to the people. It is pure cosmic reincarnation science that the image of various spiritual entities exists in the spiritual minds of the people because that is the way the cosmic reincarnation process works. In general a devout religious soul will reincarnate into a family of the same religion. The new body will be trained in the same religious concepts and will also contain the prior religious concepts in his or her soul. This insures the continuity of the particular religion throughout the ages. In effect people are born to a religious faith prior to conception.

An important interaction between the spiritual mind and the physical mind occurs when typing a book with new thoughts. Often when a writer becomes somewhat sleepy or in a trancelike state, the words coming out of the typewriter will come from the soul's mind and not the person's mind. At times the soul will take over the body to express itself. This occurs in music as well. A soul of great musical talent will take over the physical mind of a young child. The child will be a great musical protégé and amaze people. However the child has hundreds or thousands of years of musical training within the mind of his soul. This also helps to explain how a musician who is deaf can produce the most beautiful music. His soul can hear the music although his ears cannot.

Many brilliant people are the reincarnate of prior brilliant people. They study things to relearn what their spiritual mind already knows. The spiritual mind is a very high speed mind. It can lead the physical mind so that a person will not detect that his thoughts and actions are under the control of his spiritual mind. The spiritual mind belongs to God's domain and has the ability to fully control a person. However God designed the spiritual mind so that most of the time the physical mind will be the master and the spiritual mind will be the slave. This enables the physical man to have a degree of freewill.

When a person prays to God for help in solving a problem, the person enables the spiritual mind to take over. A scientist may have a difficult problem. He prays to God for an answer. Normally during his sleep, his spiritual mind will work on the problem and the answer will appear to the man when he awakens. This is normal. The person by praying to God allows the spiritual mind to take over the job of solving the problem. Of course if the prior history of the man's soul was unscientific, the soul would be of little help in solving the complex scientific problem.

In general many of the great scientific discoveries are the result of physical data entering a man through the spiritual realm. The man may not believe in God as such. He may not pray to God. However as he contemplates the problems, the data is fed into his soul. The super high speed computer intelligence of the soul within the lower mind of God calculates all possibilities while the man sleeps. The solution is then fed into the man's physical mind. The man awakens to a great discovery. The solution was the result of another man who may have died one hundred years before and who added to the overall scientific intelligence of his soul. This data was absorbed by the spiritual processor and thus was available to others.

Many people inherit somewhat evil souls who did not deserve to perish in the pit of hell. These souls are slightly more evil than good. However they still can be redeemed. A particular man whose upbringing turns him toward evil and who contains a somewhat evil soul will be reinforced in his evil actions by his soul. He will not feel compassion

35

for others. He will not feel sorrow at the damage he does to others. Often he will permit his evil soul to take over him. He then becomes a super-evil person. Data from his soul will show him his prior evil acts. He will delight in them. Thus many people in prisons have evil souls. There is little hope for them. At a certain point in their reincarnations, they will perish in the pit of hell. The Godsystem pushes evil downward and goodness upward. Some will perish upon this Earth at the end of days. Then they will turn into the collective bacteria soul in eternal chaos. When the universe fully expands the collective bacterial soul will disintegrate. As the universe disappears, only chaos at our light speed will remain

Data is also transmitted from one person to the next. A person in need of help cries out to God. The data goes from his physical mind to his spiritual mind. The spiritual mind exists all over the spiritual Earth. It is combined with all others to form the Earth God. At that point 2650 miles from this Earth, the data is available to all other souls. A visionary person who is spiritual acts as a receiver of data. The cries for help are received by the man's soul. This is then transmitted from the spiritual world to the physical world and then to the man's mind. He then goes to the aid of the person.

Sometimes a person will enter a zone of danger. An evil person awaits him. His soul will pick up the soul of the evil person. He will tell the potential victim to change course. He will say that there is danger ahead. The person will believe that God spoke to him and warned him. However this is not usually true. The usual contact we have is with our souls. It is true that our souls are part of the lower spiritual structure of God which is called the spiritual processor. Yet it is not controlled by the highest level of God. It is merely controlled by a computer program which was designed by God.

The God of the Universe is mainly concerned with Godself and the forming higher spiritual creation. The lower creation is a necessary component of the total process but no action by God is required. All we have is the spiritual processor which creates spiritual collectives that we believe to be God. We are now at the bottom of the total creation. We serve so that the higher creation can form. God has provided an upward pathway for those of us who deserve higher salvation, and a part in the forming higher creation.

In general people only remember a little bit of their past lives. The reincarnation process cleanses most of the memory of the past. In addition our present lives flow into our souls and this occupies a greater portion of our spiritual mind.

In effect the spiritual mind tends to load the new data of our present life and erase the old data. A child is a good example because his present life is short. As he lives his past slowly fades away.

Just as our minds grow old and we forget things, our spiritual mind has the same problems. There will be some salient memory of the far past, yet the immediate past will be most reachable.

The collective memory of the Earth God has the same problems. The souls from prior Earths bring intelligence to the evolving Earth God. As more data comes from present lives, the evolving Earth God drops the knowledge of the prior Universe. When the past universe is completely gone, we become the new past universe. At that time there is no memory of the far past universe. To us it appears that man started upon this Universe but that is not true. The problem is that almost all the collective memory of the past has been forgotten.

The Prophets of God tend to speak in terms that everything started here upon the present universe. Yet the astronomical data shows a rapid expansion of the universe from the original big bang. This means that a lot of prior events happened that have been erased from the collective memory.

It is important to understand that the memory within our soul is like a first in first out packaging system. New items enter the system and older items leave the system. As new memories come in, the oldest memories leave. The memory within the soul will store the memory of our past life and little memory beyond our past life.

The child will be born with a soul that contains his purified past life. The reincarnation process always purifies some of the evil of our past existence. The concept of our original sin applies to the extent that our souls are not completely purified at the time of birth. The hope of the cleaning of a new born in a body of water or a waterfall is to cleanse our souls of prior evil.

Many people hope to cleanse their souls of their present evil deeds and also of their former evil deeds from their past lives. In general the most evil of us have been erased from existence. It is possible that some people may believe they were an evil person from history. The most likely answer is that the person was a follower of that evil person but did not merit elimination from existence. Such a person carries that sin within him but he can be cleansed by having several more lives as a righteous person. Let us now look at the soul of a child as it develops.

Often a child is very religious. When he prays to God he feels his soul clearly. The soul is the closest he can get to God. As the child grows up his life feeds into his soul. His original soul at birth was his past soul. As he grows he loses some of his past soul and his present life takes over. As the child grows into an old man, his former soul has been erased and replaced by his present soul.

We start existence as a past soul and slowly over time we become a present soul. By the time we die our original soul is gone. As a child we may believe strongly in God. As we grow old we may become an atheist. Those people who devote themselves exclusively to God will maintain their souls. When people occupy themselves with completely secular activities their souls change into the soul of a secular humanist.

Souls are not eternal. They may pass from one earth to another and have many lives. Upon the same Earth they may have one hundred reincarnations. When the souls turn away from God, they no longer serve any Godly purpose. Then they perish harmlessly in death. The soul of an atheist went from a reincarnated soul in birth to a continuously shrinking spiritual entity. In death his soul disintegrates and he is no more.

The believer can say that there is a God and the atheist can say that there is no God. They are both correct. In effect the atheist becomes a purely physical man over time. The believer maintains his spiritual identity. There is no eternal punishment or hell for the atheist. He merely ceases to exist.

The term eternal soul does not really mean that a person's soul will last forever as an individual. All it means is that a person's soul can have as many as one hundred lives per Earth. The soul can then move on to the same level of Earth or higher levels of Earths. In the end of the higher process the soul takes part in collective existence. At this very high level the self is lost to the collective.

An individual soul can exist for several thousand years before being destroyed or absorbed. If it is destroyed it is nothing at all. If it is absorbed it become part of God's higher creation or Godself. In any event the eternal soul only indicates the possibility of a huge number of lives and final absorption by God. It does not mean an eternity of individual lives or individual existence.

Chapter 7: The Knowledge within the Soul

Let us look at the knowledge within the soul. As we look at a simple memory device we find that it contains a huge amount of information in a very small volume. It could easily contain an audio/video log of the important events of our lives. Typically we repeat things in our lives over and over again. Thus if one of our typical breakfasts is stored that would be sufficient together with many of our special breakfasts.

Our physical brain cleanses itself of irrelevant data. Except for a few exceptional people, we do not remember all things in our lives. The people who do remember most of their lives most likely have a strong connection between their spiritual mind and their physical mind. This gives them a glimpse at what their soul remembers.

The spiritual mind has been designed by God to remember all our past deeds. It may remember our entire life or merely the salient points of our entire life. The importance of this is that our soul will be judged. It will mentally suffer for our deeds and actions. Since the soul is comprised of many lives, it has a general moral make up and personality traits which were formed over these lives. It will suffer for our life in comparison with its moral makeup from previous lives and with a moral yardstick as defined by the human component of the God of the Earth. This level of God is the purified collective of the Gods of man.

Our tribal Gods will absorb us as long as we are aligned to our tribal laws and regulations which will vary from tribe to tribe and people to people. They will reincarnate us back to our tribe or religious group. However above this level of God is a higher level of God that can reincarnate us among other tribes or destroy us into chaos. A person may believe that he or she stands in good judgment before their own religious collective but they will not be permitted to continue existence if they have violated the moral norms of their time period.

Long ago it was perfectly acceptable to engage in immoral tribal behavior in which enemy tribes were wantonly killed for their land or women. This period of our development occurred when we were evolving from a basic tribal animal level toward higher humanity. In addition the other tribes were at the same level. The worst of tribal man were destroyed in the spiritual process but the majority of tribal men were reincarnated back into the tribe.

Over time, the standards of expected humanity increased. From tribal law the humanity of man was slowly uplifted to tribal love and compassion. The standards of the tribe were to help their fellow man. Love slowly replaced Law and the evolving Earth God slowly moved upward toward a higher level. As the humanity of man increased the humanity of the Earth God increased as well. The result was that those tribes who failed to maintain an upward increase in humanity would produce large amounts of individual tribal members who were destined for the pit of hell and erasure. Although these people attempt to obey the ancient tribal laws, their tribal God has evolved upward and the earth God has evolved even higher.

Why must the soul of man suffer for the sins of man?

The answer appears to be that the physical mind in general is the master mind and bears most of the responsibility for the deeds and actions of the physical/spiritual mind combination. However the spiritual mind influences the physical mind. It makes an individual feel guilty for certain actions. It causes an individual to have various dreams to change the behavior of the individual. Some souls are quite evil and they influence the physical mind to do terrible things. Therefore

41

the soul is partially or completely responsible for the actions of the physical person some or most of the time.

The only way the soul of man can be cleansed is by comparing its present life with the molded personality traits from its previous lives and the yardstick of the Earth God. Therefore it contains important information from its present and some information from its previous lives within its memory. In general, the soul of man refreshes the data within it, and erases the previous data. If a person remembers back 500 years, then he existed at that time and merely remained in a spiritual state until the present. It is very unlikely that a person could remember more than one life backwards unless he only lived short period of times in each of his former lives. There approximately is a one hundred year limit to the soul's memory.

In order to bring the present life into proper focus for cleansing, the soul of man will regenerate the individual. All people who are saved will come back to life out of the memory of the soul. The suffering we have caused others will be focused upon a recreation of us in the spiritual mind of our soul. In effect we must face the consequences of our actions. We will suffer mental anguish for the horrible things we have done in our lives. We will feel mental happiness for the kind things we have done in our lives.

How long does this process take?

The duration of suffering may be rather short or rather long depending upon our deeds and actions. It could take some souls a terrible amount of time. There could be an interchange of data from one soul to the other. The victim's soul could encounter the redeemable killer's soul. However most people who murder others end up in the pit of hell and are erased rapidly without any knowledge of their judgment. The Godsystem does not waste time or energy upon evil men and women. They merely return to chaos from whence they came.

Can mental pain simulate physical pain?

It is possible that the soul of man in a mental state can still simulate artificial physical pain. If the God of the Universe has designed the lower section of God's mind to simulate physical pain then the redeemable doer of evil will feel simulated pain as it stands before its own soul. It appears to me that a redeemable person who is partially responsible for the death of millions cannot feel all their pain. Yet he will feel sufficient pain in the process.

The justice system in the spiritual world is also interesting as a man or woman is brought before those he lead to do evil deeds. The low level evil leaders often took part in the murder of many other people. Each low level leader will suffer for his sins and he will also suffer from the hatred of his followers. As the soul of the low level leaders suffer for their own deeds they have to face the souls of their followers who they lead astray. This only applies to those people who sought redemption for their horrible sins and who then repented these sins for the rest of their lives.

The higher level leaders do not have to face any suffering in the world of the dead. A person only suffers in death if he or she can be redeemed. Those who cannot be redeemed will never awaken in the world of the dead. To awaken them serves no ultimate purpose. Their souls are sent directly into the pit of hell and rapidly erased. There are those who feel all evil doers should suffer for their terrible sins. However the God of the Universe is a merciful and just God. The punishment of man is left to the judgment of the Angels of God within the evolved earth God. The angels are extremely humanized entities. As such the minimization of the suffering of man at the hands of the Earth God of justice is of paramount importance. Modern civilized man puts the worst of man to death as painlessly as possible. The Angels of the Earth God will allow no worse.

The angels of the earth God are produced from the highest level of humanity of all the people in the world. This puts the final judgment of man upon the collective humanity of the human race. This insures the humanization of the evolving Earth God which like the God of the Universe is merely a spiritual machine processing center. To be judged by a machine could be quite scary but that is not the case. Within the machine is the collective memory of the highest quality level of humanity. The machine process then judges us to that standard. Of course first we are judged by our tribal or religious Gods. The Earth God insures the quality of the product.

In the cosmic reincarnation process, justice must be perfect. Therefore sufficient data must exist in the spiritual mind to insure perfect justice. We can forget our major sins but our spiritual mind cannot. In addition once we move upward from our own spiritual mind to the collective spiritual mind of the entire Earth, nearly infinite memory capacity exists. Every moment of every day of our lives could be recorded. However it is unlikely that all events will be recorded. It is most likely that salient events will be recorded. In addition the cleansing of the soul destroys most of the evil memory for those who walk on the path of God's salvation.

The entire universe in all dimensions and light speeds is very similar to a computer projection. It is a digital universe as energy flows from one point in the universe to another. Energy jumps in a digital manner rather than flowing uniformly in an analog manner. Thus the entire universe is a digital machine.

When a ball is thrown, the ball does not move uniformly from one point to another point. Every dot wave within the ball jumps to a corresponding point in space time. Some may move one space, others two spaces. There is a distribution of jump spaces so that on the average the ball jumps one space.

Can God control the jump of every dot wave?

It is unlikely that God can control every dot-wave of God's physical structure. Therefore God is a controller or programmer of the Universe. The total universe is the body of God. It is a mixture of harmony and chaos. God cannot control the entire universe. God cannot control all our actions. God cannot produce a perfect system at all levels of existence. God has Godself and the forming higher heavens close to God. This is strongly controlled by God. As we move downward toward us, more and more chaos exists.

When we mix the creative and loving intelligence of God with chaos, we produce Godly forces and anti-Godly forces. The dual forces are the result of chaos mixing with the original Godly forces. This turned them into equal and opposite forces which are embedded into space and time.

Evil then is a necessary consequence of the production of the physical world from chaos. The direct production of man and angels in the spiritual world cannot produce any evil. God for all eternity can have a spiritual Man and Woman and angels of God without the slightest bit of evil. If God was happy with this type of robotic creation then this entire universe would not have been necessary. God was not happy with his own direct creation and that is why we exist.

Although many people believe in the concept of a Kingdom of Heaven near God, such things are not possible for us at our present level of existence. The future highest angels of God are created out of man by a complex spiritual process involving billions of billions of purified individuals. The forming angel product is Godly and will exist in perfect harmony with God. There will be no chaos at that level.

The Kingdom of Heaven of the God of the Universe is but a future dream. Some of us are worthy to reach a lower Kingdom of Heaven level in our present lifetime after having lived many lifetimes. Some of us can hope for a higher Earth. Most of us will achieve this Earth or this same level of Earth.

The Kingdom of Heaven of this Earth itself is the local spiritual collective or tribal God. When a person dies, he enters his religious collective. There is a period of suffering and then he merges into the collective soul. This is a beautiful state for sure but it is only temporary. The Tribal soul can only exist as populations maintain themselves. Those people who have entered the Earthly Heaven must return to life anew. Earthly heaven is only a short period of time between lives. The reality is the continuity of physical to spiritual to physical existence until the soul is ready for higher redemption.

It is important to understand that we cannot sin against God. We can only sin against our fellow man and God's creation. Those who wantonly destroy the Earth, sin against God's creation. Those who harm animals unnecessarily sin against God's creation.

Our soul and the part of our soul that is our life will suffer mentally and possibly physically for all our sins against our fellow man and God's creation. All the animals have souls and we will have to face their souls in the world of the dead. We may very well suffer the pain of the animal that we mistreated.

Truly moral people believe that an animal such as a cow should be killed as painlessly as possible. In this manner we reduce the pain that could come back upon us. Each species of trees has a collective soul. If we destroy a forest without replanting, we cause spiritual suffering within the collective soul of the tree. In death we will feel such pain for the wanton destruction of God's creation.

Every living entity has an individual or collective soul, and every action we take causes suffering on some other soul. Even a little bug has a soul and they have a right to live. Often it is not possible to avoid killing them accidently but they should not be killed wantonly. Of course we have the right to defend ourselves against those insects which will do us harm.

When you deal with large species with collective souls, you do not cause spiritual pain to the species if you kill a few members for food. The only time that large species feel spiritual pain is when huge deaths occur such as during a forest fire. A man who wantonly sets a forest on fire must suffer the spiritual pain inflicted upon the forest's soul. This comes from the trees and all the animals that suffered death.

We came from the lower species and we must consider the pain we inflict upon them to enjoy our lives. People who wear fur coats must take into account the pain inflicted upon the animals as many were tortured in the process. The manufactures must suffer the greatest pain but those who buy the goods will suffer as well.

All of us must be very careful when we aid and assist the suffering of highly intelligent animals such as killer whales or dolphins. Once an animal or even a fish reaches a highly intelligent physical state, the same is true of the spiritual state of the animal. Killer whales often have individual souls.

In the physical world we do not readily communicate with the killer whales. We can sense their pain but they cannot tell us how great their pain is. In the spiritual world there is an interconnection between each soul and the spiritual collective or God of the Earth; which has human and non-human sections. This center of communication integrates the minds of the various species. A soul in reviewing his past life will remember the interactions with the animal world. An individual killer whale will be able to communicate with his tormentor. The man will be able to feel the pain of his animal victim. Another man will be able to feel the love of his pet dog that he cared for well.

Even the visitors to zoos and aquariums will be able to sense the pain of those in cages or pools. We must suffer the pain of our complicity in the harmful conditions we allow animals to live in. In general chickens have collective souls. As long as the chicken species maintains itself in great numbers, the death of a chicken for our food does not produce spiritual pain. We can eat chickens and cows without any suffering as long as the animals are treated humanely.

In general the pit of hell awaits the worst offenders. This is a crime against the creation and the judgment is most often spiritual death. The people who cut the fins off sharks for their soup are one example of people whose crimes against the creation are so terrible that salvation is nearly impossible. This is especially true of those on top of the economic pile who profit from such horrible activities.

Chapter 8: The Rebirth of the Soul

In the general reincarnation process, the mother is the receptor of the soul of the new born. The reincarnate enters the mother's body at various times in the production of the new born. Sometimes the soul will enter the mother prior to the formation of the fetus. The reincarnate must be compatible with the mother. A mother of one particular religious persuasion is a receptor for a reincarnate from their religious collective God.

The spiritual processor in the lower mind of God will transmit the soul into the mother. Sometimes the soul will be transmitted a few months after the fetus is forming. If the reincarnate is transmitted into the mother prior to the fetus, the mother will have two souls within her. She will have a very strong desire to have a child.

As the fetus grows a point is reached where it can accept the soul of the reincarnate. The soul is comprised of an ultra high light speed energy field that is part of God's lower mind. Once a viable child is of sufficient growth, the spiritual soul can flow into it. Initially the soul starts to form within the mother's mind. After a few months the spiritual mind is able to attach to the forming child's mind. In the process there is a transfer of the spiritual energy from the mother to the child.

When the reincarnate is part of the mother's mind, its mind tends to be in a sleep state. Sometimes it will awaken and the mother will sense the existence of another entity within herself. Often she will interpret this effect to mean that she is in communication with God. Some women may feel that they are being impregnated by God. Other women may feel that they are being impregnated by an alien.

The reincarnate can exist within the mother to be for a long time before conception. The will to return to life of the reincarnate will hasten the sperm/egg combination. It will also make the female more sexually attractive to the male. This often results in the desire of the male to have offspring. In this case, the soul of the male senses the soul of the reincarnate. The urges of the female to have offspring is often matched by similar urges within the male. The desire to have children is often more spiritual than physical.

Most tribal people in general have individual souls. They maintain their tribal identity through their tribal beliefs and practices. Most tribal people believe that they should have many offspring. When the tribe increases more than the death rate, the souls of other suitable tribal people are born into their tribe. When the tribe decreases, the souls of the tribe are reborn into other tribes.

The purity of the tribal soul increases as tribal people reincarnate within the tribe over and over again. This involves a constant population. When people reincarnate to a certain religion over and over again, the souls become pure. When the religion expands, they often bring more primitive tribal souls into their people.

The tribal laws maintain tribal soul purity. The formation of large religious collectives involves a process in which many different tribal Gods are converted into a particular religious collective. The purity of the new religious collectives depends upon hundreds of years of reincarnations into the same collective. The question for the religious group is whether it is better to have a purer religious collective soul or a larger less pure collective soul.

Many religious groups tend to maintain various ways of living in order to purify their souls. Some people attempt to live in the past in order to maintain their religious identity and the purity of their souls. Modern secular life tends to damage and often destroy the religious souls of man. The various tribal laws tend to isolate people from society at large and in so doing helps to maintain their tribal identity and their reincarnation pathway.

Modern man is often reincarnated by means of the Earth God level. Those who deserve to exist will be born again to secular or spiritual people. There people in general have no religious or little religious affiliation. Some will be absorbed by the Earth God as part of an angelic collective. Most will merely become part of the Godly salvation process where the memory of them will be erased and their spiritual energy will return to the Earth God. This will assist in the production of new souls which have no religious affiliation.

New earth salvation is in the hands of the tribal Gods. The transmission to a new Earth will be carried out by the Earth God but this level of God will merely process those who merit transmission to a new earth by the standards of their particular tribal God. In effect the Earth God has the power to do what the tribal Gods cannot do but is preprogrammed serve the purposes of these tribal Gods who have been set up to produce candidates for higher salvation. However the Earth God will destroy the souls of many tribal God candidates who do not meet the moral standards required for higher salvation. In effect each person who qualifies for higher salvation is judged by two different levels of God.

Chapter 9 Praying to God

When people pray to God, what are they really praying to?

No one can pray to the supreme entity, the God of the universe. This level of the creator is separated from the lower levels of the creation. The creation is like a beautiful picture on the wall. We see the entire picture. If we take a magnifying glass we can see the paint strokes. However we cannot see each individual dot of paint. We surely cannot see each molecule of paint. The God of the entire universe cannot see us. God has little concern for the individual. God's major concern is for the spiritual collectives.

The highest level of God has no relationship to individuals. However God has set up a complex Godsystem which reaches down to the individuals. These are lower levels in the mind of God. The lower we go, the lower the lightspeed of the energy field. The lowest level of the Godsystem is our individual souls. This is our first level of communication with the spiritual world. The majority of people never reach above this level.

The majority of people pray to their own souls. Some people manage to reach above this level to their spiritual collective or Tribal God. This is a collective of souls of a people which form the tribal Gods. The tribal Gods themselves are within the created structure of the spiritual processor. The spiritual processor is the Earth God and the tribal Gods are subsets within the Earth God. This entire structure was the product of the God of the Universe.

The souls of the living are part of their tribal Gods or religious collectives. When the people die they are processed and temporarily stored by their tribal Gods. Some of the people become angels of their tribal Gods and therefore become a permanent part of their Gods. In general when a soul in death flows into his tribal God he feels the love and warmth of becoming part of the collective entity. Then the individual will be processed and stored digitally. In the end a person's soul becomes a series of computer bits.

The angels of the tribal God form the personality and humanity of their God. In general the angels remain as part of the tribal Gods. Sometimes the angels may return to spiritual existence as they appear as dreams or visions to highly spiritual members of the tribe. In general the angel will enter the soul of the receiver. This will cause the spiritual mind of the soul to project physical images into the physical mind of the receiver. This will cause the person to believe that he or she sees an actual angel external to himself. This is because the human mind tends to recreate an image externally to it since it cannot readily conceive or visualize that the angle is actually inside the person. Later the angel will return to the tribal God and merges into the collective humanity.

The same is true of other individuals who have achieved tribal angelic status. Some of them have already moved on to higher Earths but the memory of them remains within the mind of the tribal God. Many people will then see the image of the ancient religious figures in their dreams and visions. What they see is not a living entity but a recreation of the prior living entity in the mind of the tribal God. In effect the result is a hologram type vision but it is merely a computer generated image.

Praying to the God of the universe is meaningless as far as the supreme level of God is concerned. Therefore who do we pray to? As we pray we reach out to the spiritual world. It is our own soul who hears our prayers. Praying or meditation is one form of reaching out to our inner being. For the most part our prayers reach our soul which exists at a very high lightspeed within the lower part of our tribal or religious God's mind. Sometimes our prayers will reach into our tribal God's mind. Very rarely will our

53

prayers reach into the Earth God's mind. That is as far as our prayers can go.

Our soul is a multiple life entity. Our soul is our contact with the Godsystem. It is our holy soul to whom we address our prayers. Since our soul is mostly external to us, it can be considered an alien within us. Our soul is in partnership with us. We belong to our soul and our soul belongs to us.

Our souls are external to us and are part of a spherical field which exists 2650 miles from the surface of the Earth. They cannot exist within us as such. The correct concept is that they focus upon us. We become the center of their spiritual existence. Since their light speed is extremely high the distance time wise from our souls to us is basically instantaneous. Therefore when a man encounters his soul he feels that it is inside of him.

Some people will absorb a mostly evil but redeemable soul. The evil soul will take over the man and produce havoc upon the Earth. The most evil ones will turn into the soul of bacteria in chaos. Most souls will remain at this level of Earth for several hundred years and up to a few thousand years. When man dies out in the future most of the remaining souls will degenerate downward into the bacterial soul in chaos at the center of the Earth.

If a man upon this Earth did not come from a cosmic reincarnate from another Earth, then he is an evolved soul from the human tribal collective. If he has a fresh soul from the human tribal collective then he will return from whence he came. Eventually all remaining souls upon this Earth will turn into a bacterial soul and perish into chaos. Many evil souls will return directly from human existence into bacterial existence in chaos. In general the soul either moves upward or moves to the same level of Earth or perishes eventually if no improvement occurs.

When people pray to God they may call God different names but it does not really matter because they are communicating with their souls and their tribal Gods. All the religions of man have different names for that which they pray to. These prayers and other methods satisfy an important requirement for higher salvation which is the communication with God through one's own soul.

Another important requirement for higher salvation is that of moral enlightenment. This is reasonably specified in the teachings of the various religions of man. In addition it is important to learn some indigenous religions to better understand the proper relationship of man with the animal world and the creation itself. It is not correct to believe that our only moral responsibility is with our fellow man.

It is important to understand that the God of the Earth has a very important human component but all the animal kingdoms and the vegetable kingdoms are also under the authority of the God of the Earth. If the various tribal religions or organized religions do not pay proper respect to the entire creation then the hopes and dreams of an individual is to no avail. The best of man from a humanity viewpoint who despoils the Earth or the Earth's inhabitants is destined to the pit of hell and erasure from future existence. The tribal God will praise the person but the Earth God will destroy the person. One important reason is that animals reincarnate into the higher worlds as well.

Upon the higher Earths you will not kill a cow but you will drink their milk. You will not kill a chicken but you will eat some of their eggs. The animals in turn will not eat each other as all will eat a vegetarian diet as will most humans. Anyone upon the higher Earth who harms an animal will get sick and quickly perish. The souls of higher man will quickly destroy any evil doer upon the higher Earth. The higher Earths are a different world with the

55

laws of God written upon our souls. Here we can do many sinful acts and still survive. There a major sinful act results in instantaneous death. People will be taught the higher laws of God upon the higher Earths.

A very important collective requirement for higher salvation of a religious group is that of intellectual enlightenment. Therefore it is necessary for groups of people who seek higher salvation to achieve a more scientific understanding of God and the Universe. Math and science are very important to God. Space exploration is very important to God. God needs to know the exact details of the condition of the universe from all over the galaxy and all over the entire universe at all levels of existence.

This information is necessary to insure the stability of the entire structure of the multi dimensional multi light speed universe so that God can use God's limited energy to correct and synchronize timing and deficiencies of the complex structure. This is not an individual requirement but a requirement of a group.

There is always room in the higher worlds for ordinary righteous and loving people. They serve to provide services for the religious and scientific classes upon higher and higher Earths. Yet the most important people to God are the scientific classes. Since so many people have turned to secular humanism and atheism, these people deny themselves higher salvation. They are vital to the maintenance of God's intelligence because the intelligence required to produce this Universe must be returned to God and to the higher levels of God's creation.

The religions of man must become more scientific in order to attract the scientific and intellectual classes of man.

Chapter 10 Spiritual Interactions

When a person dies, his soul seeks new life. The soul is a cell in the mind of the person's tribal God. A soul exists all over the spiritual Earth at very high light speed. The soul can be transmitted thousands of light years in seconds to new life upon an Earth at our level. If it deserves it, it can move upward one step to a higher Earth. If it does not deserve this level of salvation, it will be destroyed in the pit of hell. To transmit the soul to a new Earth, it must be acted upon by the next level of the Godsystem which is the Earth God. These levels of judgment are basically computer type processes. They do not involve the supreme God of the Universe.

In general most people reincarnate here upon this Earth. This means that their soul is active upon the sphere of the spiritual Earth. Some people are gifted in that they have strong audio communication with their own soul. They hear voices and think it comes from God. In addition fewer people have visual communication with their soul. The audio/visual communication can be considered visions.

In general the soul of man tends to absorb intelligence from the man. At times they will communicate with the man. Rarely will they try to take over the man. The proper position is man the master and soul the slave. However some evil souls take over. Then we get a monster. Often the worst of man has been taken over by his soul.

The source of the evil souls tends to be from the lesser Gods of man. Most of the Gods of man are part of processes which attempt to raise their followers to a high level of spiritual purity. Some tribal people remain at a primitive tribal level and their Gods remain at the same level. As long as the people do not perform great evils they will continue to reincarnate back to their tribe. The Earth god will not destroy the entire population of followers. Yet the worst of them will be sent to the pit of hell by the

Earth God. Over time some will be redeemed and their collective God will move upward toward greater humanity. If that does not happen then all the followers are doomed to be destroyed in the pit of hell.

Intermarriage and the secular society will often place a mostly evil tribal soul within modern civilization. If the history of the person's soul has lived up to reasonably moral norms, then the soul will reincarnate into the people of higher moral religions. This causes a mostly evil soul to become part of another people. The net result is that the transplanted soul has a chance to move upward or move downward. The result is horror that is inflicted upon the people of this world by ancient undeveloped tribal souls.

There are some spiritual people who are able to reach their own souls readily. If a grieving person comes close to them their soul can interact with the grieving person's soul. Data can be passed from the grieving person's soul to the medium's soul. From there the data goes to the medium's physical mind. By this means the medium can learn about the image of the dead in the mind of the grieving person. In effect the medium reads the mind of the grieving person. The medium can then tell the person what the dead person says. Is the medium truthful? Does the medium really understand the spiritual interaction between his or her soul and the grieving person's soul?

The medium may be innocent of being dishonest to the grieving person. The medium may also be guilty of dishonesty. In either case communication of the medium with the dead is relatively impossible except for those mediums who seek their own loved ones. A mother could communicate with a dead child. A wife could communicate with a dead husband. The reason is that the person is dead but the soul is quite alive. For a period of time after death, communication between the living person's soul and the dead person's living soul is quite possible. It may appear in dreams. For some people it may appear in visions. The interactions between the living and the dead may go on for many years. Even when the dead person's soul reincarnates into a new body, such interactions can still occur.

There is one other very interesting interaction between the living and the dead. When a person dies at home, the soul of the person can reincarnate into a loved one. A brother may die rather young in the presence of a sister. At the time of death, the sister will become aware of light flowing out of the eyes of the brother. It will be a strange sensation but the soul of the brother will enter the body of the sister. The sister will then have her own soul and her loving brother's soul within her.

The same will be true of many old people. The death at home of one may cause the soul of the departed to enter the mate. This is difficult when people die alone in hospitals. It is very important for people to die at home with their loved ones. Alternatively, the hospitals should allow the loved ones to remain by the side of the dying to enable the transfer of souls. This insures the ability of the dying person to continue to live within the body of his or her loved one.

This is not a permanent situation. There will be a time when the departed will leave the loved one and reincarnate to new life. However often the elderly couple will remain together until the other mate dies. Then they will share the spiritual world in the mind of the spiritual processor for a period of time until both reincarnate.

Physical/spiritual interactions are interesting to study. In near death experiences people will see themselves outside of their body. In this case there is a transfer of data from the eyes of the doctors and nurses to their souls. From their souls the data goes to the soul of the patient. Then it enters the mind of the patient. The intense emotional and spiritual interaction caused the patient to see himself through the eyes of others. The spiritual mind of the soul of man always looks down upon the man from a spherical perspective. The soul has no eyes as we do but it does see us by the interactions with our molecules. The mind of the soul then recreates us to match the data it gets from our brain and the brains of those around us. This accounts for the ability of a dying person to see himself outside himself.

The dying people may see the light of God. They may conceive of themselves within the Kingdom of Heaven. They may see various religious entities. The phenomenon is caused by the fact that the soul of man contains this data within its memory. Once the person is in a near death experience, the person's mind turns inward and he sees his own soul clearly. His soul exists in a collective with many other souls in a particular religious group. Then the person feels the love and comfort of the collective which is his religious or tribal God. In his mind he has reached the Kingdom of Heaven. However this is only a stop in the total cosmic reincarnation process.

Another interesting interaction occurs within a religious institution. The intensity of prayer from each person flows into their soul. This is picked up by the souls of the congregation. This produces great spiritual reinforcement. It gives each worshipper the feeling that somehow God is listening to all of them. It is merely the magnified spiritual energy of the masses that is felt by each individual soul.

As we look back to the most evil people in the history of man. We see great masses of people praising their evil leader. The individual souls were spiritually reinforced by the masses. It appears to them that the leader is a new spiritual leader. He professes hatred against the people of another tribe. Then the masses were spiritually driven to hate the other people. The net result was the degradation of the souls of the masses. The evil leader had damaged their souls so badly that most were turned into bacteria.

Today we see large religious institutions where the spiritual reinforcement is huge. It may feel good to be among the masses of mankind praying to God together. However all you get is the collective soul of the congregation. This is okay if there is also individual prayer or meditation when alone at other times. The reason is that in a quiet room you are alone with your own soul. In a huge institution how do you communicate with your own soul when thousands of other souls are there?

It is important to be a member of a religious group. It is important to be part of a spiritual collective. However it is very important to be in communication with your own soul. When you pray to God alone, your prayer goes from your physical mind to your spiritual mind. This enables your spiritual mind to be part of your life and existence. It makes you a total physical/spiritual person.

Your prayer may or may not go into the mind of the spiritual processor. That is not very important. It is very important that you know your own soul. This can only occur when you are alone.

In general self prayer to your tribal God is very important in achieving salvation. Any man can speak to their God without anyone in between. Organized religions are helpful to the masses of mankind but self-prayer to their God is in many ways superior. Often it is helpful to pray atop a mountain or beneath the stars. This puts the individual soul in private communication with God. Religious institutions only provide collective prayer. The bad point of this is that the religious leader stands between each man and his God. It is much better that each man should speak for himself.

Some people have a visionary mind. A visionary mind is a situation in which a person has a strong connection between his spiritual mind and his physical mind. In the case of various prophets of God, you could not separate their physical mind and their spiritual mind. Upon the higher earths, the connection gets much stronger. When you get to the very highest Earths there is no separation between the physical mind and the spiritual mind.

Upon this Earth for new souls the connection between these minds is very weak. The evolutionary process improves the connection after many lives. Those people who are cosmic reincarnates from lower Earths to this Earth have stronger connections than people who came from the Darwinian evolutionary path upon this Earth. Therefore some people will be very religious. Other people cannot believe that a God exists.

If a person cannot feel the spiritual dimension, he is a new evolved soul. The connections are very weak. Atheists tend to have new souls from this Earth. This limits their ability to see the truth of God and salvation. Yet some will move on the pathway of God. In time over many reincarnations they will achieve a stronger soul and a stronger connection to the spiritual world.

Chapter 11 Spiritual Evolution and Healing

Let us understand the interactions between the soul and the body when it comes to both evolution and healing. The soul is part of an ultra high light-speed energy field which is within the lower part of God's mind. It exists all over the spiritual Earth but it is focused upon the location of the physical body of a person. Within the spiritual Earth the soul is part of a collective entity. The individual soul only exists as a separate entity as part of the tribal or religious God. However some free spirits have souls that are part of social or philosophical collectives. These take the place of the religious collectives.

People conceive of the soul of man as being a part of his body. They conceive of the soul of man being confined to his body. This is not correct. The soul of man is focused upon the body but extends outward to the Earth's spiritual radius. The soul is external to the body but it is focused upon the body. In addition, the same soul can be shared with many other people here on Earth. In that case it is a collective, tribal; or family soul.

The evolution of life and man occurred because the data for man and life existed in the mind of the spiritual world and this caused man to evolve. As the bacteria or other basic life formed we had a collective soul. The soul of man originated in the soul of bacteria. Individual bacteria do not have a single soul. A blade of grass does not have a single soul. Each species of grass has a collective soul. The spiritual energy which forms the soul slowly evolves. There is a combination of Darwinian forces and spiritual forces which act together to produce life upon an Earth.

The spiritual total of life upon this Earth can be called the Earth God. It is the collective spiritual and physical center of life upon this Earth. The spiritual energy is absorbed from spiritual energy in the Godsystem. This flows into the earth and produces bacteria and basic life. As the bacteria grows and multiplies, the spiritual energy increases. Life is able to absorb the spiritual energy available in the universe.

All this is ultra high light-speed energy. It tends to be homogeneous energy which is in the form of photonic energy. It is pure mental type energy. As bacteria splits and evolves into lower life forms, the homogenous bacterial soul splits into many different forms of life. The process is heterogeneous. Singular life forms split into multiple life forms. The spiritual energy obeys laws similar to our physical laws at our light speed. The net result is that many species are produced.

When we look at a tree we do not usually find a singular soul. We have a collective soul. As the intelligence of various species develops, collective souls tend to split toward individual souls. Once that happens, the individual reincarnation process occurs. In general most homogeneous bacteria reincarnates into collective heterogeneous bacteria. When we reach the higher life forms such as an elephant, individual souls are developed. One elephant can reincarnate into another elephant.

When we look at tribal man we find that the physical body of very early pre-man evolved from prior animals. We also find that first pre-man and first pre-woman evolved to be pre-man and pre-woman because they existed in the spiritual mind of the collective soul of this Earth. First pre-man and first pre-woman did not have separate souls. Their children were born with the same common pre-human soul. Until cosmic reincarnates entered the picture, the souls of pre-man and pre-woman came from the original collective bacterial soul which evolved to be early pre-human man. Once the human cosmic reincarnates from the prior Earth entered the picture, the pre-human spiritual collective was transformed into the early human God. This process placed the soul of spiritual Man and Woman in the mind of the God of the Universe;

into the world of the original Earth and into this Earth as well.

In time, tribal man and tribal woman produced children of the tribe with a human tribal collective soul. Early tribal men and women all shared the same human tribal soul. When tribes split, different tribes developed different souls. All new tribal members had the same particular tribal collective soul. As man progressed the tribal soul split. It evolved into family souls. Over time, family souls became individual souls. However even today there are many tribal people who only have common souls.

As man developed more, men developed individual souls. Tribal souls evolved into individual souls. Once the individual soul develops, people can reincarnate back among their own people or reincarnate among other people as different people intermarry. We then get people with mixed souls. In one life they served one God and in another life they served another God. This causes the many tribes or religious groups to become spiritually connected. It also helps larger religions to take over the people of smaller religions.

When we look at spiritual healing we note that our bodies contain a physical structure and a spiritual structure. The spiritual structure is composed of God's energy. The spiritual mind contains the ability to override the physical mind. For those people who are cosmic reincarnates from other Earths, the ability of their souls to heal their bodies is very high. For ordinary evolved reincarnates there is less ability of their souls to heal their bodies. However even for these people spiritual healing is possible.

When people go into a religious institution for healing, the spiritual energy is very strong. Some members may be cosmic reincarnates. When such people touch the sick person their higher level of spiritual energy can flow into the afflicted. Then we get spiritual healing over time. In

65

general it is a slow process. When people pray for a sick person some spiritual energy flows to the sick person. The prayers of many help the healing process

There are some people who are reincarnates from a higher Earth. These people form some of the religions of man. They are very important because they will lead their followers to the higher Earths or higher levels of angelic salvation. Such people are also spiritual healers because they have excessive amounts of spiritual energy. In addition they are able to tap the spiritual energy within the Earth God.

Such spiritual healers were very important for early man. They often depended upon these individuals because there was no modern medicine or doctors to help them. The spiritual energy of the healer would flow from his soul into the soul of the afflicted. Then the energy would flow into the physical mind of the afflicted. The physical mind of the patient would then self-heal the person. If a man loses a leg or an arm spiritual healing will not help. If a man has cancer spiritual healing can help the body's immune system to destroy the cancer. There is nothing really miraculous about spiritual healing. It is just part of the interaction between spiritual energy and physical energy.

The interaction is often between the soul of the healer and the soul of others. Most times it is between the soul of a person and the body of the person. The soul can heal the body. Sometimes the spiritual energy from the collective Tribal God or earth God can flow into the patient. This would produce a miraculous cure.

Sometimes the soul is in despair at its predicament. The reincarnation process gave it a sickly body or a very unpleasant family. The spiritual mind exists in sadness. When this occurs a point is reached where the soul wants to die. It wants the present body to die so that it can be released to a new and better life. In this case, the spiritual energy of the soul acts to destroy the body. The person becomes depressed and often wants to die. Spiritual suicide is the cause of many deaths. The soul wants to die.

Lately there are a lot of people who not only want to die but want to hurt many others as well. They kill others so that they will be killed in return. This does not help them. It destroys their soul. To commit suicide under curtain circumstances is understandable. It alleviates the present pain. Depending upon the circumstances the person could still reincarnate to another life.

When the physical body exists in great pain and suffering, there is no point in continuing life. Assisted suicide is a moral thing for such people. It enables their soul to move on without the memory of a horrible death. Modern medicine can prolong a horrible death for quite a long time. It is far better to aid a person to die with peace and dignity and among loved ones. The soul can then enter a loved one before fully reincarnating into new life.

People should die near loved ones in order to enable the soul temporary sanctuary before moving on to their tribal God and then another new born. This provides a degree of comfort to the deceased and also his or her loved ones. A dying child may exist within his mother for quite a long time. The same is true of a loving spouse. Often a partially senile woman may give the impression that her husband is alive. In reality her husband's soul is alive within her. She will see him and talk to him. He in turn will talk to her. Then she will die and be together with him within their Kingdom of Heaven inside their tribal God. Many times the man will reincarnate first. The woman will follow later. Then they will be drawn together. Such is a marriage that is made in the heavens. In truth it is part of the cosmic reincarnation programming. The reason is simple. God loves lovers and he has programmed his spiritual processor accordingly.

Often the soul will suffer alone until the spiritual processor finds an available new born. Sometimes a soul must wait a long time in the lower mind of God. Then it is alone to contemplate all its life and history. A soul can suffer alone until a new body is available. This period enables the soul to purify itself to an extent. This kind of purification is not torture. No physical pain is involved but spiritual pain does occur. It is mental pain for sure. The pain is being absent from life. The pain is the inability of the soul to exert any influence over living physical matter. It is a period of time for judgment. Will the spiritual processor return the person's soul to this level of earth? Will the spiritual processor bring the soul to a higher Earth? The soul will be judged not by God but by a Godsystem which will process the soul's present life as compared to his developed traits over many lives.

If a soul is worthy of some redemption it will move on to another life if it so deserves. If a soul is not worthy of redemption, it will be destroyed. The many life soul of a person can have a death sentence imposed upon it by a computer system which was set up by God long ago. A man today can die both of the body and of the soul. He will experience physical death. Then he will experience a period of contemplation. Finally he will experience spiritual death in which his human soul returns to the level of bacteria. Most people will survive the contemplation period as even in death many can be redeemed. The worst of man will never experience this. They perish immediately upon death.

The average man or woman need not worry about such things. There is both perfect justice and perfect mercy in the Godsystem. There may be very long life but there is no eternal suffering. Spiritual death will be rather quick. The soul will merely be judged and then erased from human existence. God produced a system of true justice and mercy. The worst of man will be judged and then sent down to the level of bacteria and destroyed.

Chapter 12 Predicting the Future

The Earth contains a mixture of people with various types of souls. Most people are Earthly reincarnates. Their previous lives were spent upon this Earth. The history of their soul goes back to early human life and does not contain any information from other worlds. These people are ordinary people whose souls are growing and maturing which will enable some of them to achieve the cosmic reincarnation process now or in the future.

There are also tribal souls. Some people still have collective souls. These people are in a much lower stage of spiritual development. They tend to think in a tribal manner. They tend to hate other tribal people and enlightened people. They are closed minded and tend to act violently toward those who seek enlightenment and truth.

Finally there are three types of cosmic reincarnates who had previously lived upon other planets. There are souls who came from prior lower Earths that no long exist and who merited this level of Earth. They have been around for quite a long time. For the most part these people came to this level of Earth in more crude periods of our existence. Some have already perished in the pit of hell. Some have done well and exist today among our population. In general since the soul only remembers the immediate past life, there is no real memory of the prior Earth.

Then there are souls who came from another earth at this level of Earth. These people tend to be religious leaders, scientific scholars, great musicians, and other talented people. In their endeavors they either predict the future in some manner or create the future by their inventive genius. Since many have already lived through our period of time, they are forward looking people. They help to guide us in a way to repeat their lives again. Much of the history of the various Earths tends to be a repeat of the events of previous Earths. In general the religious history of all Earths tends to be identical since the same religious leaders appear upon all Earths. The reincarnates from these other Earths could have some memory of their past upon the other planets. However since the history of the other planets is very similar to this planet, they will tend to believe that they came from a prior life here.

The various prophets of the religions of man tend to be cosmic reincarnates from this level of Earth. They basically repeat their lives at this level of Earth over and over again. They lead their people to earthly salvation. The cosmic reincarnate establishes his God and his Heaven. When the followers die, they will find him and then reincarnate back to life.

The followers after a period of soul cleansing, will find their religious leaders as they become part of their tribal collective. This is a period of great joy. The followers believe that they are becoming part of God. They feel the love and comfort of being with their people and their God. The God of the universe has set up a system in which the individual soul gets the highest possible reward. However this condition is only temporary. Then the person must return to the cosmic reincarnation process. Most will reincarnate over and over again into their tribe. The greatest reward occurs when their spiritual energy is cleansed and returned to the God of the Earth at the angelic level.

The God of the Universe has provided both love and comfort to all peoples. Have they been deceived? The various holy books indicate that there are many Gods. People have the free will to study the different God choices. Then they can decide which God to follow or no God to follow. Some religions specify future physical life. Other religions specify angelic existence. Still other religions specify the absorption of the soul by God. The hardest paths are those who involve a multiple lifetime of struggle and devotion to their tribal God. Most often death and destruction is the price paid by those on the hard path in order to achieve the greatest reward.

Many of the Gods are easy Gods. The masses of mankind like easy Gods. As a minimum all the easy Gods bring the individual right back to where he started from. At a maximum the individual is eventually turned into a collective angel of God or absorbed by God. The easy Gods help to eliminate the individual soul from physical existence which is the goal of many. However since there is no individual spiritual existence, many other people would prefer physical existence over non individual existence.

Finally there are cosmic reincarnates from higher Earths. Most of these people exist in their time and place. They have reached a very high moral level and faithfulness to God. They lead more beautiful lives upon the higher Earths. They have no knowledge of electricity or computers. They will live an innocent life upon the higher Earth. They will have healthier bodies and live longer lives. These prophets of God came from a higher Earth and later returned to a higher Earth. They reincarnated here from a period of time where science was little known. When they returned to the higher Earth, they continued to reincarnate to the present time upon our higher new Earth. Throughout the universe copies of these same people exist in their time and place upon our level of Earth.

For millions of years our history will be reproduced over and over again. Copies of the same people will continuously come back to life to basically repeat their lives over and over again to within a high degree of accuracy. Some people feel that our lives are predetermined. In many respects that is true for society as a whole. The masses of mankind are lead by people who existed over and over again. This will cause history to repeat over and over again at our level of existence.

However as our light speed universe erases, only those universes above us will exist. The saved will move up and the unsaved will perish in chaos.

The perfection of the higher worlds depends upon a degree of predetermination here. This means that although individual freewill has a degree of variation, collective freewill is basically non-existent because those who lead society are driven by their spiritual minds to repeat their lives over and over again. There are many religious people who believe that everything that happens is determined by God. This is surely true for the prophets of God. They have no choice in the matter.

Many did not want to be a prophet of God. They resisted but they could not change their predetermined fate. Some have wrestled with angels. In truth they merely fought their own souls. The prophets of God struggled with their souls but could not succeed. They had a job to do and they could not fight their destiny. Then they returned to a higher Earth in their time and place. From there they reincarnated over and over again upon the higher Earth. They live today with very little knowledge of their past life upon this Earth. There are many duplicates of the prophets of God throughout the Universe. However that is not any concern for us since those who seek the prophets today will find the original reincarnated prophet from this earth upon the higher Earth. However they are not easy to recognize since they carry no memory of their past life upon this Earth. They will be the same basic person but the memory within their soul only lasts one reincarnation.

The history of the planet repeats all over the universe until the end of this level of the total universe. The religious leaders are called different names all over the Universe. Most will appear all over the Universe in their time and place. Their names will be different. The tribal rules and regulations they propose will be slightly different but in general it will be the same. The religious books are a fixture of the Universe and exist within God's mind in a more perfect form.

Some prophets of God are higher spiritual entities. They have existed in the Mind of the God of the Universe prior to the creation of first man upon a first earth long ago. We could consider such prophets to be spiritual sons of the Earth God. In reality they are the sons of the humanized portion of the Earth God which is man. They do not have any relationship to the creator God since this level of God is pure machine intelligence. However they did exist in the mind of the creator God prior to this Universe.

The higher Earth cosmic reincarnates will have their own history within them. Their spiritual mind will contain their own immediate past. Their physical mind will contain the present. The prophets must have a visionary mind. This enables them to readily communicate with their spiritual mind. If they had an ordinary mind, they would not have been able to communicate with their souls readily. An ordinary mind can use drugs to get them into communication with their spiritual mind. They may have dreams or visions in which communication occurs. However the superior visionary mind is able to communicate readily with their spiritual mind.

As we look back in history we find that certain prophets were able to reasonable predict the future. How is that possible? These people are certainly cosmic reincarnates who lived in a period whose history was very similar to the history of the man at that time. They most likely died in that time period. They reincarnated to this Earth in a similar time period. In general for the cosmic reincarnation process at this level of Earth, people will reincarnate to new earths in the same general time period.

The prophets from a higher earth had already lived through a similar period upon this level of Earth and were able to predict with reasonable accuracy the future events. The religious leaders would declare the man to be a prophet of God because they figured that only God could know the future. It is certain that God can reasonable predict the future or influence the future but there is a degree of chaos in the equation. The exact account of an individual's future life is unknown to God.

The Godsystem of the entire universe knows the future quite well in general because history keeps repeating over and over again with small variations. The higher Earth or similar Earth cosmic reincarnate is able to communicate with his spiritual mind. His physical mind matches present physical history with past spiritual history. The result is the ability of the cosmic reincarnate to accurately predict the future.

In some cases some cosmic reincarnates can predict the future a hundred years or more ahead. This is because they were born in a later time upon another Earth. Their soul went back in time to this Earth. This gives some visionaries the ability to predict the far future with reasonable accuracy. In truth they are not predicting our future but merely specifying their prior history.

In general a cosmic reincarnate does not lead an identical life. Chaos and random chance precludes the ability of an Earth to repeat its history accurately. When we move upward toward the Godhead, chaos and random chance is greatly reduced. At the far future Godhead there is no chaos or random chance. There the angels of God will continuously repeat in perfect precision their lives. At the highest level of physical life we will find the perfection of man and woman living perfect lives as well.

In effect it is not really desirable to strive for the highest level of existence. Such a level is quite boring. Existence there is without challenge. The highest levels of salvation are less human and more robotic. Here we have a mixture of harmony and chaos. This makes life here a challenge and somewhat fun. It is a mixture of good and evil. There is happiness and sadness. In the words of Voltaire, "This is the best of all possible worlds". Who really would want to exist in eternal perfection? Most people would not.

The continuous repetition of a perfect life may be the dream of some but the reality is hardly fun. Once we achieve the highest levels of physical existence, free will is gone. We lose our ability to be free of our souls. We live and die and repeat our lives exactly. We know our past and we know our future. Our physical bodies are driven by our spiritual minds. Chaos is gone. We are loving creatures who have independently been created out of chaos. We are collective entities comprising billions of individual souls. We know all our immediate collective lives. There is nothing new to learn.

In the end of the process, we are Man and Woman in physical paradise who knows what God knows. The process produces pure children of God who love God for all eternity. Beyond that level of existence are Man and Woman in spiritual paradise. These are composed of billions of billions of purified souls of man. The same is true of the angels of God. In effect the end products of the spiritual process are children of God in the energy of God. They love God and God loves them.

The road to this final product took an original big bang and many other cosmic events to produce. Some religions believe that God cast Man and Woman down from spiritual paradise and into the physical world. That is true but the reason is not anger but love. God wanted Man and Woman to be loving children of God. At the same time God did not want them to be robots or puppets of God. God wanted independent children of God. That is what God is producing. All of us are a little part of this spiritual process. Yet in the very end, the entire creation will return to God alone as all variations and chaos are eventually completely eliminated.

Chapter 13: The Production of New Souls

In the future, man will die out upon this Earth. Some people will inherit similar Earths and some will inherit higher Earths. In this process the soul of some of these people continues to exist. The cosmic reincarnation process insures some very long life individual souls.

Those people who did not come from the previous universe in the new earth cosmic process have newer souls. They came from the human collective spiritual energy of this Earth. Some of the souls are quite new and undeveloped. Some have experienced many lives. As we progress upward toward higher and higher Earths, no new souls will appear in the process.

This will be the last universe where human collective souls evolve into individual human souls. The remaining process will be the souls of prior universes entering higher universes. In this way the horror of the past will not be repeated in the future. The wars of man will all be gone. The many religions of man will be replaced by a singular religion for all mankind.

One of the greatest sins of man is overpopulation. The religions of man are guilty of promoting overpopulation. They want more souls for their Gods. The Tribal Gods do not need any more souls. They need higher quality reincarnated souls.

Where do the new souls come from? As man overpopulates the availability of reincarnated souls from this Earth and prior Earths is depleted. Some souls can come from elsewhere in the universe where Earths have been through the destruction of their civilizations. However this is not the usual process for the majority of mankind.

In general as a new fetus grows, a new soul will form. This is caused by the spiritual interaction between the human body and the human collective spiritual energy field within the lower level of the Earth God. This human collective energy contains very limited intelligence and purity. The refinement of the soul is caused by the reincarnation and religious processes. These new souls are basic primitive human collective souls. Overpopulation tends to bring primitive human collective souls into the population of man.

The first life of a new soul was okay long ago. Children were born into a tribe and were taught to follow the tribal rules. In this way the human collective soul was partially humanized to a tribal level. Upon death the soul was either destroyed or became part of the tribal collective. The process caused a mixing of primitive human collective souls with more pure tribal souls. The new children then had a mixture of purified tribal souls and new primitive human collective souls.

Today overpopulation among various religious people will cause impure human collective spiritual energy to enter the spiritual collectives. The children will be a mixture of more pure reincarnated souls mixed with impure human collective souls. This will cause a degeneration of the various religious collectives. Over time the various collectives will purify as the new souls reincarnate many times.

The primitive human collective spiritual energy mixed with the human collective tribal souls combine to form part primitive/part tribal collective souls. Throughout the ages, the spiritual evolutionary process has brought the primitive human spiritual collective into the human population.

The chimp/apes evolved physical structure became the basis of the human chain. The collective spiritual energy of the Earth God absorbed the DNA patterns for the structure of pre-man leading to man from information which came from the original creation process upon the prior Earth. Over a very long period of time the human level of God formed within the mind of the Earth God. The selection of a particular suitable chimp/ape animal by the evolving human God caused the children of the selected animal to have the DNA patterns of pre-man.

Once pre-man was formed by the spiritual process, the primitive human God gained in intelligence. Over a period of one hundred thousand years or more, refined DNA codes were inserted into successive stages of pre-man leading toward man. The stage was set for the insertion of the DNA patterns of Man and Woman into pre-man.

The spiritual data within the soul of Man and Woman from the prior Earth went into pre-man and caused man and woman to be born upon this Earth. The same is true upon all other Earths in the Universe. The bodies were our bodies but the data within their souls was very limited. They were born ignorant and innocent. The human level of the Earth God at that time was also ignorant and innocent. All that had happened was that the DNA codes of man were transferred from the God of the Universe to the God of the galaxy then to the God of the sun and finally the God of the Earth.

The process produced innocent human beings. The same basic process occurred upon each Earth in the present Universe. First Man and first Woman here had innocent children. They had to learn how to survive. They had to learn how to protect themselves against animals.

They did not stay innocent very long. Cosmic reincarnates from the prior Earth at their primitive level were transmitted by the prior Earth God into the present Earth God. This increased the intelligence of the present Earth God dramatically. The children born to first man and first woman were cosmic reincarnates. Within the memory of these primitive reincarnated children were the means of survival.

The children had children and civilized man took shape. The process cleansed the reincarnates of most of their sins and their memories. They knew how to hunt but they could not remember hunting upon the previous Earth. At that time they did not have individual souls but had collective souls.

The transfer of souls from the prior Earth to the present Earth is timed to prevent advanced souls from entering the present life. The transfer of the souls of previous religious man to this Earth involved huge amounts of time. The souls could have been transmitted earlier but the transmission is digital data. The data contains the DNA codes of the individuals and complex codes which determine their traits and personality. In addition their purified history and other intelligence are transmitted as well.

This data is converted by the Earth God and a soul is ready for birth within the human God. In this way the children and grandchildren of the original Man and Woman were brought upward in human intelligence. In time the homogeneous human God transforms into the Gods of man. It is these Gods of man who rise to the highest levels of intelligence. This in turn raises the level of intelligence of the Earth God.

At the same time the human level of the Earth God remains a source of spiritual energy and new souls. Some of this energy flows into the forming Gods of man. Other energy flows directly into forming fetuses. This causes some children to be born with more developed souls from tribal Gods while other children are born with less developed souls from the human God. However the Earth God also processes some soul which exists at a very high spiritual level.

In addition, the souls of cosmic reincarnates continue to enter the picture in their time and place. The net result is a complex mixture of collective tribal souls, individual souls from the tribes, primitive souls from the lowest level of the human God, and cosmic reincarnates from the prior Earth. To this are added the cosmic reincarnates from higher Earths who bring with them knowledge of the God of the Universe.

Chapter 14 Communication with the Earth God

Once we reach above the tribal Gods we reach the God of the Earth. This higher level of the Godsystem processes all life upon the Earth. It is connected to the collective of the human tribal God souls. It is also connected to the collective of animal souls as well. This is the highest level of god that we encounter here upon this Earth.

This level of spiritual intelligence has the capability of transmitting the souls of man to new Earths. It processes some of us during the earthly reincarnation process since not all people are connected to the many religious and tribal Gods. Many religions have many branches and each branch occupies a separate section within the Earth God. Often these sections vary upon national boundaries as people of the same faith often hate and war with similar people of the same faith but with different national or ethnic boundaries.

Two different people with the same religious books often war with their neighbors. They each believe that God is on their side but in reality their Gods are different. Large religions once combined smaller tribal Gods into one larger God but in time this one God became many different versions of that God. The net result was that each group occupies a different section or level within the mind of the Earth God.

In general the Earth God does not take part in the particular religions of man. The religions of man reincarnate people into their religious group. Those people who are not affiliated with any religious group fall under the control of the Earth God. This God will reincarnate people wherever it chooses to find a vacant spot. In most cases those souls who have no affiliation with a religious group will end up in the overpopulated sections of the Earth. Many of the souls who do not merit any level of salvation will merely be destroyed by the Earth God.

The Earth God will also select the best homes for those secular humanists who serve the Earth quite well. They will reincarnate into those families who have no real religious affiliation but who serve the Earth God in the protection and preservation of the Earth. These people may not pray to God but everyday their work and deeds serve God best. Such people can also achieve higher new earth salvation or angelic salvation upon the authority of the Earth God who never speaks to man. Yet this God can radiate people and heal them. In many respects this is a level of God that is Love whereas all the tribal Gods produce tribal laws and are therefore Gods of Laws. Laws can be spoken to man but love can only be demonstrated.

Once we move above the earth God level we reach the sun God. Many ancient people used to worship the sun God. This was a period in the history of man when people reached into their souls to find the source of their existence. As we look at our souls we find that they are connected to the Earth God.

The Earth God is the spiritual interaction of God's photonic energy with the physical energy of the Earth. This level of God processes us and prepares us for transmission to other Earths at our level or higher Earths. In order to do this our souls must be digitally transmitted from the Earth God level to the Sun God level. There are no other viable planets within our solar system. Our data must be retransmitted from the Sun God level to the Galaxy God level at the center of our Galaxy.

This level of God is necessary in order for our souls to travel from one planet to the next. Once we go above the galaxy God we enter the God of the present Universe. This level of God is necessary to transmit us from one galaxy to the other.

There is a level of God of the past universe and a level of God of the present universe. Finally when we go above the three universes, we reach the level of Godself. This level of God is separated from us. We cannot pray to this level of God.

Our first communication with God is our souls. This is the lowest level of God. This is followed by our tribal Gods. These are our spiritual collectives. Once we reach above these two levels of God we find the Earth God. These levels of God are what concern us.

The level of God that is the sun God plays a very important function in the elimination of life upon this Earth. The original creation process copies the Universe of the prior past unto this present universe. This means that long ago a form of our sun and our solar system existed. Very early forms of life also existed.

I do not know how many prior universes there were. We can assume that the prior past universe was the start of the present universe. We can assume that man first appeared upon the past universe. Therefore within the probability function, many Earths with man upon them came into existence in the present universe.

There are millions of earths such as ours. Chaos will produce some that do not produce human life. They will not develop or they will be destroyed by chaos. Therefore God does not have to use God's present energy to create a universe. The only thing that God has to do is to transport the souls of the saved to new Earths.

The important function that requires the effort of God is the total destruction of a solar system and a planet Earth. Once the Earth has yielded a harvest of souls, there is no necessity for the Earth to remain.

The future for mankind is quite horrible. There will be atomic wars. Man will be consumed by flesh eating bacteria. It is very easy for God to pull the plug on this solar system. That is the importance of the Sun God. If God pulls back the interconnecting godparticles or godphotons between the physical and the spiritual worlds, gravity will fail. The sun will very rapidly expand into a huge fireball. It only takes a little over eight minutes for the radiation to hit the Earth.

This burst will kill all life facing the sun immediately. The fireball will take a little longer. The people on the opposite side of the Earth will be killed off within 12 hours as the Earth rotates into the radiation. Some people buried deep underground will live until the fireball reaches the Earth. Therefore some people may actually live a few days. Once the fireball hits, the entire Earth will be destroyed.

Of course if at that time God released the Earth God level, then this Earth would explode in a few seconds. It may be that God will release the sun first to absorb the saved souls remaining upon this Earth. Whether we die instantly or slowly over a few days will be determined by God.

In any event we see in the universe areas of space time that are empty. We see stars that have exploded. In some areas the universe is still creating new stars and new planets. Some saved souls may go there. Other saved souls will go on to the future universe.

I do not like to predict the future. Someday it will come and I do not want to know when. It could be years from now or centuries from now. One thing is certain, the future will come.

Chapter 15: The Multi-Light-Speed Universe

In order to understand the cosmic reincarnation process, it is necessary to understand the multi-light-speed/multi-dimensional Universe that we live in. In 1981 I started the study of space and time after I became aware that the universe we live in was a multi-light speed universe. It is composed of multiple universes of increasing light speed energy up to light speed infinity at the Godhead. The God of the Universe exists at the highest level. It is also important to understand that at the highest level, our entire universe at our light speed is no larger time-wise than a single human mind.

As I studied all the different possibilities for the structure of God and the Universe, it became clear to me that the entire universe was composed of plus and minus dot-waves. Over the years I have attempted to understand what that meant. Although the complexity of multi light speed multi dimensional space time is for great mathematical minds, I will present an Engineering viewpoint of the subject. This is only an approximation to the complex reality involved. The true solution is a combination of Einsteinian space time, quantum mechanics, and string theory to which Doppler Space Time, dot-wave theory, and multi light speed theory is added. When you add this all together, the complexity is really impossible for the human mind to comprehend. Therefore the best I can present is a simplified ballpark version of reality.

Let us begin with a simple version of a dot-wave. A dot wave can be considered a wave that exists at a single point or focal point within a simple sphere of a particular radius. The radius is much larger than the radius of the Earth. Most of the atoms upon this Earth are composed of dot waves of the same radius. When we go to the solar system we find that dot-waves extend from the center of the sun outward to a distance far from the sun. The galaxy produces even longer dot-waves extending beyond the galaxy. In effect the patterns of the dot-waves mirror the gravitational field of the Earth, the Sun, and the Galaxy. Once we go beyond the galaxy the dot-wave patterns get more complex. This gets us to Einsteinian space time and the more complex Doppler Space Time.

Einsteinian space time is the best mathematical fit to Doppler Space Time. Einstein's equations are the root mean square of similar Doppler Equations. However this is the best we can do since the Doppler Equations require Fourier series solutions which are much more complicated. The only advantage of Doppler equations is that the clock paradox does not exist. However since Einstein's equations work so well, it is best to say that his equations are a best first order approximation to space time reality.

Although the entire universe is composed of only two basic things, we have to add string theory to understand why things are so complex. Space time is quantized into approximately thirteen dimensions according to string theory. This means that the focal point of some dot-waves jump from one corresponding group of dimensions to other corresponding groups of dimensions. The dimensions they jump to and the path that they take determines their characteristics. A simple sub-particle will be composed of billions of dot-waves.

Once we add quantum mechanics to the story, we find that there is a probability that large groups of dot-waves will occupy certain dimensions most of the time. When we add sub-particles together we get very complex patterns of multi dimensional dot-waves that produce an electron. The electron will exist at a certain location most of the time. To make matters more complex, the electron could exist at other locations some of the time. Therefore you cannot depend upon the existence of an electron at any point in space time.

The reason for this is that the center of the dot-waves is merely focal points of the wave. The wave itself exists external to the focal point. For this Earth, the dot-waves exist thousands of miles from the center of the Earth. The focal points can hop all over the place. A particular electron can be in New York in one second and Chicago the next second.

When we look inside our bodies for something, we will find nothing at all. We think that we have something inside of us but all we are is a huge number of focal points. At the dot-wave level there is no such thing as mass. The property of mass is a reaction between huge numbers of focal points and surrounding huge numbers of focal points. The only reason we exist is that the probability of a large group of focal points remaining at the same location is basically one hundred percent. We cannot guarantee that a single electron will remain where it is. However we can guarantee that a huge number of electrons will remain where they are.

In the same light the probability of being able to transport a human being to another location is basically zero. An electron can jump thousands of miles at the speed of light but a block of iron cannot. We see that quantum mechanics and string theory when added to dot-wave theory gives us an understanding of our basic universe. Now it is necessary to understand basic Einsteinian space time. The question is what makes time?

The God of the Universe is eternal. God does not know of time in God's domain. In effect the past, present, and future are all one in the mind of God. We experience time. If you take our universe in the time dimensions you will see that we have expanded in time. The Universe of today is larger than the universe of yesterday. What happened to yesterday? What will happen tomorrow?

We then get back to how God made time. We see that the universe is expanding and that yesterday has been erased. Tomorrow is forming. This means that there are actually three universes. This causes continuous creation. The universe of yesterday dies out and the universe of tomorrow forms. Since everything in the universe is quantized, the formation of the new universe occurs in steps. These steps are very small but they are discrete. We can call this time a split second. This split second is the basic clock of the universe. It is the physical distance between the universe of the next split second and the universe right now divided by the light speed.

This causes the universe to have a sub-atomic clock. Our atomic clocks are a large scale version of the clock of the universe. Although space time is much more complex we can look at the total physical universe as a spherical sandwich. Although the universe is changing by quantum mechanics, at any split second we find three universes. We exist on the surface of the present spherical universe. Closer to the center is the previous universe which is slowly erasing. Further from the center is the future universe which is forming.

When God compressed the dot-waves which existed in chaos into a small ball, it started out as universe number 1 which is the past and universe number 2 which is the present. Universe number 1 was the initial conditions. The transfer of energy goes from universe number 1 to universe number 3. This means that as the past is erased the future is forming.

If the energy moved from universe number 1 to universe number 2, we would not have a very stable existence. God tied the past and the future together to insure that the present could support life. This sandwich of three physical universes gives us the universe that we see and measure. The clock runs one way. We cannot go back in time upon this present universe. However there is always the possibility that we could travel to the previous universe which is being erased. It is also possible that we could travel to the future universe that is forming. I do not believe we can do this but the possibility exists. The big problem is that our dot-waves are tied to the universe that we live in.

If we were independent entities that actually had substance within us, such travel would be possible. Since we have nothing inside of us, we get our existence external from us. This means that only God can transport us from the past into the present or future. Our souls exist within God's ultra high light speed photonic energy. Our souls can travel from one universe to the other but our bodies cannot. Due to the dot-wave theory, we are limited to this present universe. Sad to say but my sci-fi novel is just a fantasy and not a reality.

The erasure of a universe occurs when individual dot-waves of the photons, electrons, and protons leave the patterns of the photons and sub-particles. The end result is dark energy which is composed of dot-waves in chaos. The universe started with dot-waves in chaos and in the end returns to dot-waves in chaos. We then get dark energy and dark matter. Everything disintegrates into chaos.

When we move up in light speed to the photonic energy fields we find that some of our sub particles such as godparticles react with the photonic fields of God. Therefore some dimensions are common to both God's dimensions and our physical dimensions. This enables God to mold the dot-waves in chaos into a physical universe.

God is a multi-light-speed entity. The highest layers make up the mind of God. The lower layers make up the body of God. Our souls are part of the lower mind of God. Just as we exist outside of our present location, the same is true of our souls. Our souls are interactions between our physical body and God's spiritual photonic energy.

Our souls exist outside our bodies but are focused upon our bodies.

The only difference between body and soul is that our bodies are composed of dot-waves that were molded into shape and form. Our souls are composed of higher light speed dot-waves that belong to God. The net effect is that our physical brain looks outward at the universe from our center while our soul looks inward upon us. The physics is similar but the direction of mental focus is opposite. Our physical mind looks outward while the mind of our soul looks inward.

The physical world consists of stars and planets while the spiritual world consists of photonic energies. Since each occupies many exclusive dimensions they coexist without interference. They also occupy some inclusive dimensions. This enables higher light speed photonic energy to shape the lower light speed energy. In general the state of the lower light speed energy is dot-waves in chaos. In the beginning we had dot-waves in chaos. This is like God's clay. God can mold the clay into universes.

The universes are only temporary. The physical world cannot remain physical forever. God's energy can mold it but in the end, it must return to chaos. Therefore physical existence is dynamic and temporary. The only existence that is permanent is spiritual existence. Therefore God can produce a spiritual world which is beyond time.

It appears to us that our universe started 13.7 billion years ago. This would appear to indicate that the original big bang occurred 13.7 billion years ago by our time clock. This is projecting time based upon our measurements. During this time, the universe has expanded and changed from each past universe into corresponding present universes. Scientists feel that there was an expansionary period in the universe. In reality the universe steps from one universe to another. All we see is the present universe and we get a glimpse of the past universe.

It then becomes very difficult to ascertain when the original big bang occurred. It appears to me that to the original we have to add multiple smaller big bangs as galaxies form. This gives us an original big bang and a huge number of distributed big bangs. Black holes will form in the present universe and explode in a future universe. All along the way the entire structure of past, present, and future will slowly erase and the energy will be returned to God.

In any event we have a very complex multi light speed, multi dimensional universe which operates upon dot-wave theory, Einsteinian space time, quantum mechanics, and string theory. All I can do is look at the complexity from an Engineering viewpoint. Let us now look at space time from another viewpoint.

The latest astronomical data shows that the universe is expanding faster than the speed of light. How is that possible?

Space itself is composed of the photonic energy of God. This energy goes to light speed infinity .The present universe is a quantized light speed energy field. Therefore for this existence we have a light speed of 186,000 miles per second. The universe of the past had a lower light speed and the universe of the future has a higher light speed. They are all fixed light speeds. Therefore the universe steps between a lower light speed and a higher light speed.

The driving force of the universes is the photonic energy of God. This energy can go down to light speed zero. Therefore we can have an original universe of nearly zero light speed. This would occur at an original big bang. The driving force is God and as God brings his muscle energy upward, the prior universe is erased and the future universe is created.

Although the physical universes are quantized, God may not be. Therefore the light speed can change in a linear fashion in God's dimensions. On the other hand if God's energy is also quantized then the driving force of the universe would also be quantized. However many small steps will still appear linear to us.

The importance of the astronomical discovery of a faster than light expansion is that as our universe erases, the energy is transferred to a second future universe. Physical energy always jumps from the past universe to the future universe. Therefore our physical energy will jump to the second future universe.

Spiritual energy always jumps to the future universe. The souls of the saved will be transmitted into the next universe which is almost finished forming. When the past universe is completely erased then the future universe will be completed physically. Of course once it is completed it will start to erase.

The most important thing is that even though our universe is light speed 186,000 miles per second, the expansion is not limited to that speed. Since souls are transmitted at the rate of 1,000 light years per second according to my visions, there is plenty of room for many future universes if God drives the universe at a constant rate.

Since God can control the rate of expansion, the driving force can eliminate the physical universe at any future time. In any event space itself appears empty but that is where God lives. The expansion of space time is not caused by the exploding galaxies. It is caused by the spiritual driving force. It is caused by God.

In order to better understand the triple universe, let us assume that we can build a space ship which could cross the time barrier. There are many difficulties in doing this but let us understand the physics behind it.

Einstein felt if we could go faster than the speed of light we would enter the past. This is not correct since the driving force is moving upward toward higher and higher light speeds. Therefore if we were in a spaceship that could reach light speed, we would move into the future if we could exceed the speed of light. At the same time if we were able to reach this point and then lower the light speed of our spaceship we would reach the universe of yesterday.

What are the Engineering problems involved in such flights. First we need to overcome the Doppler length problem. A spaceship would get shorter in the front and longer in the back. The distortion would kill everyone involved. Let's assume that we could produce a force field to overcome that problem. Then we need an engine that could produce enough power to bring us up close to light speed. This would involve breaking apart the proton and producing photonic energy from it. Let assume that we could do this as well.

Now we have to leave the present universe. Right now as we are upon the Earth, our dot-waves are part of the Earth's field. Each of us is very large in size. When we leave the Earth, our waves tend to stretch outward to become part of the sun's field. The saving grace is that as we move faster and faster, our waves tend to shorten in the forward direction.   If we can decrease our forward dimension to a tiny size then we are in the position where we can pass into the past or future universe. A burst of forward energy will propel our dot-waves into the future.

This energy changes our physical structure into higher light speed energy. We then enter the future. On the other hand a burst of energy in the reverse direction will send us into the past. In general our souls will remain the same because they are composed of ultra high light speed photonic energy.

What will we find? The past universe still exists. It is erasing but some animal life will still be there. It is doubtful that we will find any human life. If we move into the future we will find planets that are forming. Some will have early man. Others will have some of our relatives. We will not find wars and people killing each other. The bodies of saved man and women will be improved. Their connections between body and soul will be stronger. The evil that exists here will not be found.

We will not find rich or poor on these planets. Houses will not have locks on their doors. You will find no jails and security people are only there to assist people in case of accidents or medical problems. People will live long lives and die only to be reborn over and over again. The body perishes but the soul is perpetual.

In general only God can bring us to the future. Yet it is possible that man can break the time barrier. God may permit man to move to the past. The brave astronauts will most likely die in the past. However God can still bring their souls back to this level of existence or even bring their souls into the future.

It is also possible that God could permit them to move physically into the future. However they would have to be cleansed of the evil of their prior existence. They would land upon a higher Earth with no knowledge of whom they were and where they came from. The people will know that they came from the past. They would be welcome but their bodies would not last as long as the inhabitants of the higher Earths. Yet when they die many will be reborn upon the new Earth.

In any event although the Einstein equations show an object getting smaller and smaller as you reach toward the speed of light, this is only true for the Einsteinian equations which are the root mean square of the Doppler equations. The front length of the wave reaches toward zero but the rear length of the wave reaches toward infinity. When we multiply the almost zero and the almost infinity, and then take the square root; we still get Einstein's solution. Therefore Einstein's answers are always excellent.

It is unfortunate that an actual spaceship will turn into primary sub-particles and photons and then disintegrate. The misuse of quantum mechanics helps some people to believe that you could actually get a spaceship through the time barrier. The probability of bringing one sub-particle through the barrier is reasonably high. The probability of bringing a spaceship through the barrier is basically zero. However it is good to study such things as they give us a better understanding of the past, present, and future universes. Sci-fi gives us the opportunity to explore amazing possibilities. As we move to the far future things will get even more amazing. Our novels can help us to understand such things.

Chapter 16: The Source of our Existence

What is the source of our existence? That is the question. How do we exist? That is another question. These are very difficult questions. We use the word God and wonder what God means. We look at the physical world and wonder how we exist within this world.

The scientists and mathematicians produce laws which seem to indicate what is inside of us and how the subcomponents of us interact with the rest of the universe. The various laws we find are little bits and pieces of measureable reality. We base our laws on what we can see and measure. Those things that we cannot see we use instruments that react with them to provide us with a picture of what they appear to be.

We build large machines such as at CERN to split atoms and produce sub-particles, matter and antimatter. Our mathematicians find many additional dimensions from a mathematical perspective. Einsteinian space time provides us with other properties of the universe from a mathematical perspective.

A grand theory of everything is not easy to produce. The mathematics is much too hard for us to understand. Will we ever be able to formulate a complete theory? Can a super-computer finally solve the complex issue? What does the super computer do? How large would it have to be to compare to the mind of God?

The mind of God is the largest super-computer that can ever exist. God's memory reaches to infinite light speed. God's memory capacity reaches to infinity. What is God composed of? How is it possible to write equations for God and the Universe that have any real meaning? It looks like an impossible task.

From a scientific and mathematical perspective the answers appear impossible to solve. We are dealing with a superior intelligence that is so complex that we really have no ability to formulate. The only thing we can do is to produce Engineering models of God and the Universe. The Engineer is like an artist. The Engineering solution is not a perfect scientific solution. It is merely a method in which the complexity of God and the Universe can be simplified to a practical level so that people can understand the greater complexity of God.

The Dot-wave theory is not intended to be a perfect scientific and mathematical solution to the structure of God and the Universe. The Engineer is taught models of various things so that he can build things that work. The pyramids were built by Engineers. Everything of practical importance was built by Engineers. We can say that God is a super Engineer. God builds universes out of the mind of God. Is it necessary for God to have exact formulas for everything God builds?

Perhaps God cannot even write equations for what God creates. The mind of God thinks things and the body of God turns forms of energy in chaos into actual things. We would like to have exact formulas for the process. However it is possible that even God does not know how to adequate explain and formulate the process.

There are several important principles in dealing with how we exist. Some philosophers believe that because we think we exist. Due to the spiritual/physical interaction, it is possible that our minds are driven by our spiritual mind. This would mean that we are under the control of our spiritual dimension. However God is in control of our spiritual dimension. One possibility is that we do not really think, and our brains merely take external data from our souls and process it. This would make us mere robots. Some religions believe that God is in control of everything. This would cause all events to be predetermined. Then we would again be mere robots.

There most likely is a combination of predetermination and free will. Some of the religious entities upon this Earth and all over the entire universe will repeat everywhere. This means that some people are predetermined. Since the history of all planet Earths are basically the same everywhere, history is basically predetermined with small random chance variations. In effect the predetermination is mixed with chaos.

If an Earth goes too far off course and does not produce proper yields of souls, then the entire harvest will be destroyed. There are millions of Earths in the process and those few Earths that do not live up to expectations will merely be erased from the cosmic process.

Earths are like fruit on a tree. Most of the fruit will be good. Some of the fruit will be bad. God will select the good fruit and let the bad fruit fall to the ground and rot. The bad fruit does not go to waste. The spiritual energy merely is cleansed and recycled.

Some of us will find the spiritual forces too great to resist. They will follow a path that God has designed for them. God's purpose will be done. Their free will becomes limited to what they will eat for breakfast and other ordinary decisions. Yet they will be driven to do God's work by dreams and visions that they cannot control.

The various prophets of God are driven by their spiritual dimensions. They are not free. They are merely robotic entities which are driven to perform whatever tasks God has assigned them. Some people believe that such people are struggling with angels. However they are really struggling with their own souls. The spiritual dimension is very strong and physical man has little to no capability to resist their souls.

To make matters worse, their souls are connected to their tribal Gods and their tribal Gods are connected to the God of the Earth. The connections go further as the Earth God level is connected to the Sun God level which is connected to the Galaxy God level which is connected to the God of the Universe.

Although the God of the Universe is too far above us, Gods influence is felt at all the lower levels of God's existence. The entire structure exists in the mind of God. Since the entire structure exists in the mind of god, we too exist in the mind of God. This produces a very difficult problem.

The simple atheistic solution looks at our existence as independent entities. We are composed of independent particles. Our atoms and everything we see and touch are made up of particles. It certainly appears that way. It looks as if we exist where we are.

The problem was later understood that sometimes things appear as particles and at other times they appear as waves. We then get mathematical theories of particle-waves. This produces fancy wave equations which seem to work out.

If we are really made up of waves, the source of our existence can be the mind of God. Suddenly we do not really exist where we are. We only appear to exist where we are. Our true existence is external to us. The Dot-wave theory looks at the physics of the universe from an inverse perspective.

A particle appears to exist in the here and now. Our measurements indicate that. However if everything exists in the mind of God, the particle exists external to where it appears. If the particle exists external to where we measure it to be, then what exists at that point? For the dot-wave theory, a dot exists at that point. The dot is a focal point of a wave.

The dot-wave theory turns everything inside out. Infinity becomes zero and huge distances become very small distances. In many respects this is similar to film development. When we develop film we produce a negative image. Then when we print the film we get a positive image. It is also similar to a foundry mold. We make a negative image and then cast a positive image.

When we look at things from a negative or inverted perspective we are looking at things from the mind of God. The dot-wave theory looks at things from the mind of God and not our mind. Things look different from the mind of God then our mind. We see and measure things from our bodies and our instruments. We look at two hard objects hitting each other and see physical objects which obey certain measured laws. The laws appear simple to write and understand. When we look at the inverted image in the mind of God, the complexity of the laws is extremely difficult to formulate.

The great electro-physicist Thevenin came up with simple rules and regulations which changed complex mathematical solutions into simple models that produced the same results. When we look at the Universe from the perspective of the mind of God, there are simple models which provide us with information from an inverted perspective.

The dot-wave theory is not a normal theory from our perspective. It does not match the laboratory results from CERN and other institutions. All our theories are from our physical perspective and our physical instruments.

The second part of this book is devoted to the Dot-wave theory and Doppler Space Time. These theories are an Engineering effort to produce models of God and the Universe from a Godly perspective. The best we can do in this regard is an Engineering perspective because only God could produce a scientific perspective. Even God may not be able to fully formulate God's perspective because what God thinks becomes reality.

Chapter 17: Initial Concepts of the dot-wave theory

The universe we live in is composed of combinations of two basic things called dot-waves. There are positive dot waves and negative dot waves. In addition there is a spectrum of dot-waves at a huge amount of light speeds from near light speed zero to near light speed infinity.

The dot-waves are purely electrical type waves but they have different characteristics depending upon which dimensions they operate in. In String theory mathematicians have found at least thirteen dimensions. Such things are very difficult to comprehend because we see three dimensions and have a feeling for a time dimension but that is about all our physical minds can visualize readily.

When the dot waves spin or travel between various dimensions they give the characteristics of mass and energy. They also appear as the electrical property of positive charge, negative charge or zero charge. In other dimensions they give the appearance of photons. Dot-waves appearing as mass less photons when operating in one set of dimensions appear as mass when transformed into another set of dimensions. The dot-waves have not changed but they have shifted into another configuration.

An important aspect of dot-waves is that they do not exist at the point where they appear to exist. They exist externally to a focal point. There is no actual substance at the focal point. We appear to exist in the here and now. A rock appears to exist in the here and now. However if you looked inside the rock for something you would find nothing at all. It only appears to be something because everything else appears to be something.

The source of our existence is external to us. We could argue that we exist in the mind of God. For all practical purposes it appears to us that the substance of our existence is where we are. When we move into the spiritual world, it is clearer to us that our soul exists in the mind of God and is external to us but centered upon us. However our physical body has the same physical structure although the light speeds are different. The soul is ultra higher light speed energy while our body is composed of energy at our measured light speed.

The universe we live in is composed of a sandwich of three universes. There is the universe of the past, the universe of the present, and the universe of the future such that as the universe of the past erases, the universe of the future forms.

There is a continuous flow of energy between these three universes. Over time the past erases and the energy of the past is transmitted into the future. However this is a slow process whereas the flow from past to present to future is continuous with only a slight amount of energy net flow from the past to the future.

The dot-waves are charged waves of very small charges. Positive dot-waves are compatible with both positive and negative dot-waves. For a given light speed all dot-waves are identical except for charge. Particles or sub-particles are composed of various amounts of both plus and minus dot-waves. A positive sub-particle will have an excess of positive dot-waves. A neutral sub-particle will have an equal number of plus and minus dot-waves. Finally a negative sub-particle will have an excessive amount of negative dot-waves.

As we add dot-waves together, the energy of the sub-particle increases and its wavelength decreases. What we see is the focal point of the wave. A single dot-wave will have a huge focal point equal to the radius of the universe. A proton which is composed of a huge number of dot-waves will have a focal point or wavelength equal to $1.321E\text{-}15$ meters. To us it appears that the proton is this tiny size but in reality it is as large as the radius of the universe.

Since the proton really does not exist at the measured location, it is easy to understand why quantum mechanics specifies that a particle could be at one location in the world in one split second and thousands of miles away in another split second. The reason is that it was never really at any location at any time. It only appeared to be at a particular location because that is where the focus of the energy was. The dot-wave theory specifies that the source of the dot-wave is external to the point of measurement.

In the same light, one of the forces of binding energy is due to the interaction between the dimensions of the present universe and the corresponding dimensions of the past universe and the previous universe. As the previous universe slowly erases, it tends to expand us. At the same time we are erasing and we push against the future universe. This causes a back pressure on matter which tends to hold mass together and at the same time causes one mass to be bound to another mass.

This makes it difficult to find a particular sub-particle or force which produces binding energy because this force is inherent in the basic design of the three universes. We coexist with the past and future universes but our measuring instruments cannot go into their dimensions. There would only be one different dimension whereas the X, Y, and Z dimensions would be the same. However the time dimension would be different. T-past, T-present, and T-future prevents any measurement of a different universe from taking place.

When we move upward to the highest light speeds in God's domain we find that some photonic dot-waves are compatible with us. Some of our sub-particles share dimensions with God's dimensions. This enables the photonic energy of God to operate upon this physical Earth. In particular God can compress a spectrum of lower light speed dot-waves into a small ball.

When God releases God's forces, the three universes will be set up. God can then operate upon the first universe to set up the basis of galaxies. This will cause the image to flow into the second universe. As the second universe forms, the first universe starts to erase. This will cause energy to flow into the third universe.

Over many cycles man will come into existence. This will occur over millions of planets such as this Earth. There is a lot of chaos and random chance in the process. The physical process basically reincarnates an Earth to a higher level each time. Man will appear and reincarnate to a higher level as well.

Once the universe was set up by God, nothing else had to be done. The process turns a spectrum of dot-waves in chaos into planets and particles. As each layer of the universe expands and erases, higher light speed dot-waves in chaos turn into order while lower dot-waves in order turn back into chaos. Slowly the universe moves upward to higher and higher light speeds. The planets get more perfect and man gets more perfect.

The best of man continues on and the worst of man are turned back into chaos. Over time the mass decreases as the light speed increases for the same amount of energy. Since God exists at the highest light speeds, man reaches a Godly state of being. In the end of the process man in photonic energy becomes part of God and eventually is absorbed by God. In the end, God is alone in the universe and most likely will restart the process from the beginning.

The dot-waves of God extend outward to infinity but are focused upon the center of our universe. The center of our dot-waves is also centered at the center of our universe. It appears that the universe is a simple sphere surface with a center such that every point in the universe is basically equidistant from the center of the universe. The plane of the universe has a thickness which involves some fancy mathematical space time equations. However from an Engineering viewpoint it can be considered a simple expanding balloon with a degree of thickness.

The previous universe is closer to the center and the future universe is further away form the center. The light speed of the prior universe is lower and the light speed of the future universe is higher. Therefore the dot-waves are quantized as per light speeds. Our light speed appears to be constant. The driving force of God may be digitized or analog. It has the capability to bring the universe down to almost light speed zero. Therefore at some prior time the universe at minus infinity had zero light speed at the pinpoint.

The three universes cause the expansion of the universe to appear in steps. We look back in time and believe that the original big bang occurred 13.7 billion years ago. This is based upon our light speed and our astronomical measurements. It is possible that the universe expanded as an exponential function. When we look back in time we only see a linear approximation to the time of the pinpoint.

This does not matter much because we can only judge things by our measurements. The most likely solution was that an original big bang started the ball rolling. Then over time galaxies came and went. This caused a large amount of distributed little galaxy bangs all over the universe. Since past energy flowed into the present as well, we get the appearance of an expansion period of the universe rather than a simple big bang. However there were many prior universes since the original big bang. We are just one level of God's creation.

In the first part of this book we looked at God and the Universe in terms of simple words to describe the complexity involved in the dot-wave structure of the universe.

Although the dot-waves are easy to speak about, the mathematics of the dot-waves and the three universes is beyond my ability to formulate. In this regard I rely upon simple Engineering models of God and the Universe to bring things to light so that an ordinary educated person could fairly understand the principles involved. The models are not perfect. They are used to bring the concepts of the dot-waves to the reader.

First we will look at a simplified Engineering model of a universe which brings the Dot-wave concepts into focus. This is a modified repeat of the dot-wave theory proposed in Doppler Space Time and repeated in Aliens Within Us. The equations would have to be modified to account for the three universes but such an undertaking would take many years to accomplish. The best I can do in my limited time left upon this Earth is to modify the previous model.

As an example to help understand this phenomenon of the past, the present, and the future, let us look at the Bohr orbit fine constant.

The equation for the fine constant is:

$$A^* = hC/ \ 2 \ pi \ K \ QQ = 137.036 \qquad (17\text{-}1)$$

Where h= Plank's constant of 6.62608E-34, C= speed of light of 2.99792E8, pi = 3.14159, K= Coulomb's constant of 8.89756E9, and Q = charge of 1.6021E-19.

The reason for the 137.036 instead of the prime number 137 is because the Bohr Orbit moves 274 half waves every 360 degrees in one plane and 274 half waves per 360 degrees in the orthogonal plane. These are not pure sine waves but modulated DC or half sine waves. For stability, we must phase lock on one-half wave shape increments as we move from one plane horizontally to the next plane below that.

The angular shift between planes is:

Angle = 360 degrees/ 274 = 1.31386 degrees (17-2)

There is a phase shift of 1.3186 degrees between each half wave. The cosine of the angle is:

Cosine 1.3186 degrees = 0.999737   (17-3)

The inverse of the cosine of 1.3186 degrees is:

1/ Cosine 1.3186 degrees = 1.000263 (17-4)

Therefore:

137/Cosine 1.3186 degrees = 137.036

We see that the fine constant correction factor is due to the fact that each complete Bohr orbit is not exactly the same. They are displaced one half wave shape from each other. The Bohr orbit is always rotating. There are exactly 137 cycles in the Bohr orbit but there is a net phase shift between one complete electrical cycle and the next complete cycle.

In standard electrical theory the cosine of the angle always applies to forces or power. In the Bohr orbit there will be a binding force between every segment of the present orbital configuration and corresponding past and future segments.

In addition the Bohr orbit starts in the past and moves to the present. Eventually it achieves the future. However the future is identical with the past and therefore the Bohr orbit is a simple example of the workings of the three universes.

Chapter 18: The Dot-wave theory

Section 18-0 Introduction

In this chapter, the Dot-Wave theory will be introduced and a model of the universe will be presented. This will enable us to understand the basic structure of God and the Universe from the perspective of the mind of God. This involves looking at the source of existence of the physical world as being in the mind of God. The theory presented here is a modified version of the theory presented in "Doppler Space Time" in the year 2000.

When we look for the basic equations of the Universe from a Godly perspective, God will see certain simple relationships as God looks at the universe. When we look at the universe our physics is very complicated. If we look at the equations of Einsteinian Space Time, Quantum Mechanics, and String theory we get very complex mathematics such that very few people could understand.

From an Engineering perspective things are less complicated. The basic solutions will involve simple algebraic equations and various constants. These equations are simple enough to understand but they all require various constants which are not that easy to find and understand.

In particle physics a rock is composed of particles in which the source of its existence is at the location of the rock. When we look at a rock from the perspective of being in the mind of God, the source of the existence of a rock is external to the rock. Therefore if we look inside a rock for something we will find nothing at all.

The universe from God's perspective is composed of dot-waves where the dot is merely a focal point of a field and the energy is external to the point of the dot but it is focused upon the dot. This produces an inverted image of how we view reality. In this respect infinity becomes zero and zero become infinity.

It is a premise of the dot-wave theory that everything exists in the mind of God. The equations presented here are an attempt to produce an Engineering model of reality from a Godly perspective. We begin with the postulates of the Dot theory.

Section 18-1: Postulates of the Dot-wave theory

The Dot-wave theory states that the entire universe is a huge multi dimensional multi light speed electromagnetic field. It states that everything in the universe is electrical in nature, and that mass is an electrical property of charge, space, and time. The Dot theory states that everything in the universe is composed of a multitude of only two types of things for each light speed dimension. At each light speed, there is a multiplicity of plus dots and minus dots. The electron, the proton, the neutron, and the photon are all composed of dots. The same is true of all the subparticles. The only difference between things is the particular geometric configuration of each particle or photon.

The Dot-wave theory states that the electron is composed of minus dots equal to the total charge Q= 1.602E-19 coulombs. The proton contains approximately 1836 times as many dots as the electron but has an excess of positive verses negative dots equal to the charge Q= 1.602E-19 coulombs.

The dot-wave is a quantum of the electromagnetic field. It is an electro-photon or charged photon. The dot can be considered a focal point of the multi dimensional multi light speed electromagnetic field.

The basis of the Dot theory is that the universe is a complex electromagnetic field. The dot charge, and or the dot current will drop as the universe expands. Likewise, when the universe originally contracted, the dot charge, and or the dot current increased. In the process, the universe expanded from a small space-time radius to a very large space-time radius.

The first effort of the book is to quantize the dots. A method has been chosen based upon a simple model universe, which has interesting Thevenin equivalent characteristics. The equivalents are used to approximate an Einsteinian space-time universe. In general, for the Einsteinian universe, every point is the center of the universe. For the Dot theory model universe, a corollary to the Einsteinian solution is that the universe looks like a perfect sphere to every dot-wave in the universe. This permits easy calculations of the dot characteristics since the dot sees a very simple universe of radius R_U.

In the Dot-wave theory, the model is used to produce electrical equations for the rest mass/energy of the electro-photon dot-wave. The dot-wave reaches to the radius of the universe moving at the speed of light there. At that point it is a light sphere. The dot-wave also exists as a focal point within photons and particles. The dot-wave always expands at the speed of light but the focal point will form standing wave patterns within matter.

The dot can be looked at as a current flow as it expands outward to the radius of the universe. This light sphere is centered upon the surface of our universe. Therefore the light sphere reaches to twice the radius of the universe and at the same time the center of the universe. By adding all the spheres together we get a picture of what the universe looks like. For the most part the universe is a spherical surface of distance Ru from the center but it contains some thickness and nonlinearity's. For the dot-wave theory we will only look at a simple sphere to represent the dot-waves and their configuration.

The dot energy will be quantized and the number of dots per neutron will be calculated. The radius of the universe and the time of universe is readily found by the Dot theory and matched to the astronomical data using simple electrical equations of an expanding hydrogen atom. In addition numeric analysis is used and seeks simple equations, which match the universe. The numeric theory states that the universe is tied together algebraically by simple expressions and simple numbers such as 0.866, 1.414, 4, π, e, 137, etc. It is important to understand how God created this universe and the dot-wave theory is an engineering analysis of reality.

In general standard electrical and mechanical equations and data are used throughout the book. The Dot theory provides the quantization of the energy of the universe. The Dot-wave theory equations also provide the basis for the conversion from Mass to Charge in chapter 19.

Section 18-2: The Model Universe

Let us start out with some very simple algebraic equations as we search for an understanding of the relationships between the electrical world and the mechanical world of the universe. In order to understand how the universe functions, it is necessary to start out with a simple model of the universe and then explore the model and expand upon it until it matches the actual measurements obtained in the laboratory. The universe is modeled upon huge numbers of expanding light spheres.

For our light speed, every point in the universe is equidistant from the center of the universe. All individual electromagnetic spheres with the center at a dot form a total complex spherical shape such that all spheres completely fill all the spaces on the spherical surface of the universe. The spheres all interlock in three dimensions of distance such that the universe always looks like an absolutely perfect sphere of radius 2Ru when viewed from outside the universe. However from any point in the universe, the universe always looks like a sphere of radius Ru with respect to that point.

To understand this we can look at the universe as a spherical surface. We live upon a plane at a distance Ru from the center of the total universe. If we draw a circle of radius Ru on a piece of paper this will represent a two dimensional view of the universe. We live on the circumference of the circle. We can then take a compass and set its radius to Ru. After that we can draw a huge number of circles of radius Ru all around the circumference. The net result is the formation of another complex circular structure of twice the radius as the original circle.

Every point on the original circle is basically at the center of its own circle. All these circles form a dark spot in the center of the total universe. The net result is that every point in the universe is equidistant from the absolute center of the universe and the outer circumference of the circle produced by the huge number of circles centered upon the original circle. This is the way the universe looks to the mind of God. Once we add the three time dimensions and then the many other dimensions the complexity increases greatly. In effect this is the structure of the mind of God. Although the universes has steps in light speed from light speed zero up toward light speed infinity, to us it looks like a constant light speed universe. Therefore for this analysis we will look at a constant light speed solution which is what the universe appears to us at the present time.

Let us look at a single sphere of light expanding from a pinpoint as representing the electromagnetic universe. As the light expands, the radius of the universe also expands and a standard ruler and a standard time clock expand as well. Time clocks slow down and rulers stretch as the universe expands. Notice that if the universe expands at light speed, it expands much faster when it is small then when it is large. The rate of expansion or the acceleration of the light sphere is:

$$A_U = (C^2)/R_U \qquad\qquad (18\text{-}1)$$

Equation 18-1 states that an expanding sphere of electromagnetic light energy has acceleration, A_U of meters/second^2 that is simply the speed of light, C squared divided by the radius, R_U of the light sphere. An idealized universe at every point follows the form of equation (18-1)

110

This simple equation forms the basis of an idealized universe comprised of a complex huge electromagnetic field, which experiences time and distance elongation.

The total force acting upon the model universe is simply the mass of the Universe M_U times the acceleration of the universe A_U:

$$F_U = M_U A_U = (M_U C^2)/ R_U \qquad (18\text{-}2)$$

When the universe was contracted by God, the dot-waves existed in chaos and resisted in a manner similar to the general gas law. Dot-waves in chaos therefore produce a pressure which resists compression.

Once the pressure from God during the original big bang was released the dot-waves exploded and at the same time sub-particles and particles were produced. At that time the electrons and protons were formed. The net effect was to turn dot-waves in chaos into intelligent structures; therefore chaos was turned into order by the compression of God's spiritual energy upon the universe.

At full compression the mass of the universe approached infinity. In this book we look at things from both an electrical and a mechanical viewpoint. The universe can be described either from electrical equations or mechanical equations.

From a mechanical viewpoint, the force resisting the expanding universe is the loss of mass per unit time. From an electrical viewpoint, this force is the loss of charge per unit time.

From a mechanical viewpoint:

$$F_U = -C \ d(M_U)/ \ d(T_U) \qquad (18\text{-}3)$$

In equation 18-3 we see that the force acting upon an expanding electromagnetic universe is the light speed C times the derivative of the mass of the universe with respect to the time of the universe. Notice that the expanding bubble of the electromagnetic field has less and less force acting upon it as it expands more and more toward infinity since the mass is decreasing toward zero. Correspondingly, the charge is decreasing to zero as well.

The radius of the universe R_U is related to the time of the

111

Universe by the following:

$$R_U = C \, T_U \qquad\qquad\qquad\qquad (18\text{-}4)$$

The radius of the universe is simply the speed of light times the time of the universe since the pinpoint at big bang. The pinpoint is not a singularity. From the outside it may appear as a single point. From the inside it is the same as the number of dots in the universe, thus it is a multiplicity. In many respects it could be looked at as a huge collection of general gas type atoms squeezed together. Many dot waves will combine at the zero point but they will never converge into a single focal point of all dot waves. God will release the pressure prior to that condition. If God kept the pressure on too long, only a single galaxy would exist in the universe today.

The energy of the model universe is simply the mass of the universe times the speed of light squared. It is also the force operating upon the universe times the radius of the universe. Thus:

$$E_U = F_U \, R_U = M_U \, C^2 \qquad\qquad (18\text{-}5)$$

Notice that all these equations are very simple general equations, which describe an expanding light sphere. Yet, they will produce reasonable results when the concepts are applied to the standard equations of physics for the various calculations shown in this book. The purpose of this study is to produce idealized equations which look at how God sees the universe. These will never exactly match the way we see the universe exactly although things will be similar.

If we set equation (18-3) equal to equation (18-2) and solve using equation (18-4) we get:

$$M_U \, C^2 / R_U = -C \, d(M_U)/d(T_U) = - C^2 \, d(M_U)/ \, d(R_U) \qquad (18\text{-}6)$$

In addition since the net or differential driving function is the loss of mass as the universe expands:

$$M_U \, R_U = \text{Constant} \qquad \text{(Kilogram Meters)} \qquad (18\text{-}7)$$

112

One solution to Equations 18-6 and 18-7 is an exponential function.

This simple solution to an expanding light sphere is

$$M_U = M_O e^{-x} \qquad\qquad (18\text{-}8)$$

$$T_U = T_O e^{x} \qquad\qquad (18\text{-}9)$$

$$R_U = R_O c^{x} \qquad\qquad (18\text{-}10)$$

$$C_U = C_O \qquad\qquad (18\text{-}11)$$

Where X is a driving function which varies from minus infinity to plus infinity.

In equation (18-8) we see that the mass of the universe decreases as the driving function (x) of the universe increases. In equation (18-9) we see that the time of the universe expands as the driving function (x) of the universe increases. Finally in equation (18-10) we see that the distance of the universe expands as the driving function of the universe expands. Thus as the light sphere increases, both the ruler and the clock expand for a constant light speed universe as per equation (18-11). The driving function can be a simple linear function. It could also be a more complex function.

Since both charge and mass decrease as the universe expands, the following equations also apply:

$$Q = Q_O e^{-x} \qquad\qquad (18\text{-}12)$$

In equation 18-12 we see that charge decreases as an ordinary exponential. It gets smaller as both time and distance increase.

$$U_O = U_{Oo} e^{2x} \qquad\qquad (18\text{-}13)$$

In equation 18-13, the electrical permeability U_O increases to the second power as both time and distance increase.

$$K = K_O e^{2x} \qquad\qquad (18\text{-}14)$$

In equation 18-14, the Coulomb's constant K increases to the second power as both time and distance increase. Likewise for the electrical permittivity:

$$\varepsilon_O = 1/ \; K \; = \varepsilon_{O_0} \; e^{-2x} \qquad (18\text{-}15)$$

In equation 18-15 we see that the electrical permittivity constant ε_O decreases as the universe expands. The speed of light is:

$$C = 1 \; / \; (\; U_O \; \varepsilon_O \;)^{1/2} \qquad (18\text{-}16)$$

In equation 18-16 we see that the speed of light remains constant as the universe expands since the electrical permeability constant increases while the electrical permittivity constant decreases. The impedance of the universe is:

$$Z_U = Z_{U_0} \; e^{2x} \qquad (18\text{-}17)$$

In equation 18-17 we see that as the universe goes toward infinity, the impedance of the universe also goes toward infinity and it becomes an open electrical circuit. When the original universe was compressed toward a pinpoint at minus infinity, the impedance of the forming universe was zero. It went from compressed chaos to become an electrical short circuit.

From this we see that the universe is an inductive/capacitive structure which varies from a short circuit at big bang to an open circuit at infinity. This is one particular model. Things get more complicated when the light speed is varied from zero toward infinity. There are many variations to this solution and each one can take years of study.

From these simple relationships of an expanding light sphere, a model of the universe can be presented. It is necessary to add to the model specific constants, which will produce values that are close to what we see and measure.

Section 18-3: The Gravitational Constant

In this section we will take a very brief look at the gravitational constant operating upon the expanding electromagnetic sphere of light.

The mass of the universe decreases with the distance the universe occupies as postulated in section 18-2. As the universe expands there is a gravitational force due to the loss of mass with increased distance. This is also due to the loss of charge as distance expands. Basically the gravitational force can be described as Ampere's law of current loops. As the universe expands galaxies, stars, and planets will form due to the gravitational forces from the expanding universe and the interaction between the universes of the past, the present, and the future. In addition there are additional compressive forces due to the interaction of God's spiritual energy and the physical universe. The gravitational force produces the effect that matter will attract matter according to some very basic equation of the universe.

There is an equivalent of the universe which defines the gravitational constant in simple terms. Let us now formulate some simple equations to yield the mass of the model universe. For an ordinary balance of mechanical forces, the force of attraction between two bodies would be:

$$G M M/ R^2 = MV^2 / R \qquad (18\text{-}18)$$

In equation 18-18 we see that the gravitational force is equal to the centrifugal force in an ordinary planetary motion equation. When we consider the entire universe to be an expanding interlocking set of a huge number of light spheres of electromagnetic energy, a Thevenin equivalent of the universe can be produced. A simple force equivalent of the set of light spheres would be the force of two universes separated by a distance equal to the radius of the universe. This simple equation appears to be how the universe works. It is not something which can be proven but it appears to be a basic equation that God would use.

We can now equate the force of expansion with the gravitational force of the universe. Although the universe is extremely complex it will behave in a simple manner. Thus:

$$F_U = M_U (C^2)/ R_U = G M_U M_U / R_U^2 \qquad (18\text{-}19)$$

In equation (18-19) we see that the gravitational forces acting upon the entire universe depends upon the mass, M_U of the universe and the speed of light, C squared divided by the radius, R_U of the universe. What is happening is that as the universe expands the universe is erasing and turning into chaos. This causes the material universe to turn into dark energy. In addition the past universe has already turned mostly into dark energy. All this is due to the loss of mass in the universe.

On the right side of the equation we have a force of the gravitational attraction of the mass of the universe with itself at a separation of Ru. On the left side of the equation we have a force of rotation of the mass of the universe around a center point of radius Ru. This equation is fine but we do not have a constant which relates this force to measured quantities from ordinary physics upon this Earth. In general electrical equations and mechanical equations tend to be vector equations. There is an angle between the forces in equation 18-19. A thirty degree angle is a common angle as is sixty degrees. Therefore a vector correction factor of 0.866 is possible in this equation. This factor will be used to produce the time of the universe in equation 18-39.

Solving for G:

$$G = (C^2)R_U / M_U \tag{18-20}$$

In equation 18-20 we see that the gravitational constant G equals the speed of light squared times the radius, R_U of the universe and divided by the mass, M_U of the universe. As the light speed increases from one universe to another, the gravitational constant would increase. At the same time the mass of the universe would decrease.

At the big bang the radius was small and the light speed approached zero. At the same time the mass of the universe approached infinity. This caused the gravitational constant at the big bang to be zero. In effect at the pinpoint nothing holds the universe together. Once God releases God's compressive force the universe will explode. It is also interesting in that as the light speed moves upward to infinity, the radius moves upward to infinity and the mass heads toward zero. Therefore at the end of the universe mass no longer exists.

When we look at the universe from a constant light speed model, the radius of the universe expands with time and the mass of the universe decreases with distance, the gravitational constant G increases as the square of the radius of the universe. Since the mass decreases, the term GMM is a constant except for nonlinearities.

If we look at equation 18-19 however, we see that the gravitational force decreases as the square of the distance with time since the mass of the universe decreases while the radius of the universe increases. Thus at full expansion, the gravitational forces weaken as the square of the radius of the universe. Therefore, stars and planets will explode. The protons will do no better since they experience a slowing of their internal oscillations. Thus protons and electrons experience their own red shift effect, as do photons. They will be destroyed as we head toward maximum expansion. This is called the little bang where the entire universe returns to electro-photon energy in chaos.

In general, the atomic binding forces, which depend upon the DC Charge Q, weaken as time goes by. This causes non- radioactive atoms to become more and more radioactive as time goes by. Right after the big bang many atoms existed which do not exist today. As time goes by more and more atoms become radioactive. Eventually lead will become radioactive and in the future even oxygen will become radioactive. In the end of the process, the protons and electrons are also destroyed. Thus at full expansion everything is destroyed. At full expansion, we only have a complex electromagnetic field left and this field will disintegrate into chaos.

In this section the formula for the mass of the idealized universe was computed using a Thevenin equivalent of the universe. Various other methods will be used to produce idealized formulas to calculate the number of dots in the neutron and in the universe, etc. These methods serve to produce a physically realizable universe from which the protons, electrons, neutrons, photons, gravity, and space-time can be understood. The whole purpose of this study is to find the means by which the universe was created.

Section 18-4: The Dot-wave structure of the Universe

In this brief section, let us understand the very simple nature of the universe. The entire universe is composed of only two different things. The first is a plus dot and the second is a minus dot. According to later calculations in this chapter, there are 1.09628E125 dots per universe. These calculations are based upon a normalized universe and everything is accordingly normalized. As more data is analyzed, corrections to the figures are possible in the future. However the workings of the universe do not change if corrections are applied. The changes are the dots per neutron, the radius of the universe, and the mass of the universe.

There are two different dots, which could be considered the focal points of the electromagnetic fields of the universe. Each dot has a plus charge, Q_D or a minus charge, $-Q_D$. Clouds of dots make up the charge Q=1.602E-19 coulombs. Each dot has a local DC charge of 2.78622E-61 coulombs, which will be calculated later in this chapter. The dot also has a charge of 1.602E-19 coulombs at the radius of the universe. Therefore it is zero size and almost infinite size simultaneously. This is due to the Doppler effect of a charge moving at the speed of light C which is explained in chapter 20.

The local charge Q of a proton is made up of large numbers of both plus and minus dots with an excess amount of positive dots. The same is true for the positive subparticles as well. The electron is composed of only negative dots in the ground state. As photons are added to the electron, the composition of the electron has a balanced blend of plus and minus dots together with the original amount of minus dots. Photons are always balanced blends of plus and minus dots.

In addition to the local charge Q_D of the dot, there is a local dot current flow I_D. There are mechanical properties of moving groups of dots and space itself, which cause mass, inertia, and gravity. In effect each dot has a local energy E_D, an equivalent rest mass M_D, a charge Q_D, a current I_D, a capacitance C_D, and an inductance L_D.

The charge Q_D with its equivalent rest mass M_D and current I_D comprise the entire structure of the universe. It is a purely electrical universe and everything is made from the dots. The dots can be considered multidimensional Doppler space-time entities.

The dot currents can represent current flow between positive and negative levels of the universe. They are perfect point current sources. On each level of the universe, they appear to come from nowhere and go nowhere. The plus dot current can be said to flow toward us and the minus dot current can be said to flow away from us. However, it is all one sandwich of a positive/negative universe.

In general, the dots exist basically uniformly everywhere and are felt everywhere up to the radius of the universe. They are little bits and pieces of the discharging electromagnetic field itself. The universe is a multiplicity of dual electromagnetic fields and the dots are the lowest active quanta of charge/energy of the field itself.

At the level of the dots, only rest mass exists. The dots are thus electro-photons or DC current flows. They have no mass in themselves but three dimensional gyroscopic patterns of dots in motion produce steady state forces, which give rise to mass and inertia. The ordinary photons moving at the speed of light do not possess a three dimensional gyroscopic image and thus appear mass-less. The proton and electron have repetitive gyroscopic patterns within themselves and produce synchronous images of their past, their present, and their future. They have both mass and inertia.

The dot is an electro-photon and is free to form mass or to form photons. The only real difference is that the photons do not produce a repetitive relatively stationary three dimensional gyroscopic image, and therefore do not obey Ampere's law of forces for current loops, which requires repetitive images. In addition photons occupy different combinations of dimensions as per String theory concepts. Another important difference between the photon and matter is that for matter its forward Doppler length is greater than zero while for the photon its forward Doppler length is zero. This prevents the photon from having any ability to form a three dimensional gyroscopic pattern.

Matter is the product of dot-waves which form patterns at a particular location and a particular time. Photons are the product of patterns of dot-waves which travel at the speed of light. They are merely different properties of the same thing.

The two lead balls in the laboratory are pushed together because of the expansion of the universe and the loss of mass per unit time. The expanding electromagnetic field of each electron, proton, and neutron, plus all the hydrogen Bohr orbits and higher more complex atomic Bohr orbits produce current loops which create a magnetic type force directed toward the center of all atoms and particles.

Although the underlying cause of the gravitational field is the loss of mass per unit time, this tends to appear common mode to us. We cannot readily detect the fact that the light speed is increasing and the mass of the universe is decreasing. The only thing we can readily detect is the fact that the hydrogen atom is expanding and this causes electrical attractions between two hydrogen atoms as will be shown in the next chapter.

Photons travel at light speed when free but drop speed slightly when in a strong gravitational field, which increases the electric permittivity and magnetic permeability constants. The speed of light is slightly higher between galaxies than within a galaxy or near a star. This causes the light to bend around the star, as the photon becomes part photon part mass. In addition space itself bends due to the interaction of the physical world and the spiritual dimension around the star.

There are many alternate explanations in the dot-wave theory. The universe can be looked on from an electrical perspective or a mechanical perspective or a combined electrical/mechanical perspective. The models presented here give the reader a glimpse into the choices that God has in how the universe will operate.

Section 18-5: The Mass of a Dot

In this section, we will look at the relationships between the electrical world and the mechanical world in a very simple manner. The charge Q will be used and explored as the source of the conversion. There will be a direct connection between the dot charge Q_D and the dot rest mass M_D.

In Section 18-2, it was postulated that the mass of the universe decreases as the radius of the universe increases. This is a premise of the Dot theory. The universe is composed of a large number of dots of local rest mass M_D and local charge Q_D. The dot also has the charge Q at the radius of the universe. The mass and charge of each dot is also decreasing as the universe expands. There is a DC current flow out of the charge Q at R_U similar to that of a parallel plate capacitor where the plates are expanding. In addition to the strong electrical coulomb DC forces acting in the universe, there are also strong magnetic forces acting in the universe.

Let us look at the DC electrical field of a standard expanded charge Q at the surface of the model universe which is a perfect sphere moving at the speed of light. We find locally that:

$$V_D = K \, Q/R_U \qquad\qquad (18\text{-}21)$$

In equation (18-21) we see that the voltage of a dot here caused by a charge Q located at the radius of the universe is very small but not zero. We have a finite universe of radius R_U, and there is a little voltage left over at the dot, which is the center of the dot's own universe. The dot looks like the charge +Q or –Q at the radius R_U. Likewise as we look at each positive charge Q of the proton, there is a little positive voltage left over at the radius of the universe. For the electron, a minus voltage also reaches the radius of the universe.

The universe is balanced with the charge Q at R_U and at our particles as well. We then have an electric battery for the universe where the current flows through the dots.

The outer sphere of the universe at radius $2R_U$ is a place where the outer surfaces of all the dot-waves form a sphere which has zero net charge but where the is no dot current flow. The same is true at the exact center of the universe. The surface of our universe where we live is a neutral conducting plane. This is where all the action happens. This is where all the dot currents are flowing.

Further out, the individual dot-wave spheres or planes are expanding. The end of the dot waves could be ordinary spheres or circular planes. Both configurations will produce the same overall pattern of the universe.

Let us now look at the dot current flow. This dot current flow is not an ordinary current flow such as with an electron. It is the slow discharge of space-time similar to the decrease of charge of a parallel plate capacitor when the plates are slowly moved apart. In effect, the dot current is the discharge of space itself. The dot current flow is the flow across the barrier between positive dots and negative dots.

The universe is a space-time capacitor, which charges when it is compressed and then discharges as it expands. As the universe reaches toward infinity or maximum expansion, the dot charge reaches near zero and everything the dots produce is reduced to a subliminal structure which is a state of chaos. As long as God maintains higher universes beyond ours, our dot-waves cannot reach a state of pure nothingness. For that to occur, Godself would have to reach a state of pure nothingness at infinity. This is not a possibility as at the extreme light speeds going to infinity, standing wave patterns exist which contain a permanent memory structure of the mind of God.

The net result is that we can turn into chaos but not pure nothingness since we cannot expand beyond the radius where the permanent structure of God exists.

We can relate the dot current to the dot voltage by the impedance of space, which is the intrinsic impedance of electrical theory. The impedance of the universe Z_U is

$$Z_U = 4 \pi K / C \qquad\qquad (18\text{-}22)$$

$$Z_U = 4 \pi K T_U / R_U \qquad\qquad (18\text{-}23)$$

where K is the Coulomb's constant, T_U is the time of the universe, and R_U is the radius of the universe.

In order for the voltage at the neutral conducting sphere at the outer radius of the universe to be zero, there must be a dot current flow equal to the dot voltage divided by the impedance of the universe.

122

$$I_D = Q/ (4 \pi T_U) \qquad (18\text{-}24)$$

The charge Q could be thought of as the peak of a very tall mountain that exists at the radius of the universe. Over a very long period of time the distance to the mountain increases toward infinity and the peak of the mountain disappears. This is caused by a current flow. The current flow is similar to ordinary electrical capacitive circuits in which the plates are separated, and the charge on both plates' decrease. It is V_D $[d(C_D)/d(t)]$ where the change of capacitance (C_D) is related to the time of the universe.

The current flow can also be expressed as:

$$I_D = QC/ (4 \pi R_U) \qquad (18\text{-}25)$$

In Equation 18-25 we have the charge Q at the distance R_U, moving away at the speed of light and causing the spherical current I_D. The current flow at the distance R_U is identical with the current flow at the local dot itself. The dot power flow into the local short circuit is the dot voltage times the dot current. Thus:

$$P_D = V_D I_D = K Q Q/ (4 \pi R_U T_U) \qquad (18\text{-}26)$$

Likewise the dot power is the square of the dot current times the impedance of the universe.

$$P_D = (I_D)^2 Z_U \qquad (18\text{-}27)$$

$$P_D = K Q Q C/ (4 \pi R_U{}^2) \qquad (18\text{-}28)$$

The dot energy is the dot power times the time of the universe:

$$E_D = K Q Q/ (4 \pi R_U) \qquad (18\text{-}29)$$

Finally the dot mass is the dot energy divided by the speed of light squared.

$$M_D = K Q Q/ (4 \pi R_U C^2) \qquad (18\text{-}30)$$

Equation (18-30) gives us the basic conversion equations from mass to charge velocity in a very simplified way. The only things necessary are any constants that may be required. This method permits us to look at the mass of a dot M_D and compare it to the coulomb constant K, the charge Q, the radius (R_U) of the universe, and the speed of light, C. It is an equation which matches the mechanical world to the electrical world at the units of kilograms. Mass is the motion of charges with the units of coulombs meters per second as will be seen in Chapter 19. Therefore kilograms equal coulombs meters per second.

The force acting on a dot is the energy of the dot divided by the radius of the universe. Thus:

$$F_D = E_D / R_U \qquad (18\text{-}31)$$

$$F_D = KQQ/(4 \pi R_U^2) \qquad (18\text{-}32)$$

We can now relate the mass of the dot to the charge of a dot by means of the magnetic permeability constant U_0 of free space. Using (ε_0) for the electrical permittivity constant we get:

$$\varepsilon_0 = 1/ (4 \pi K) \qquad (18\text{-}33)$$

$$C = 1/ (\varepsilon_0 U_0)^{1/2} \qquad (18\text{-}34)$$

$$K = (U_0 C^2) / 4 \pi \qquad (18\text{-}35)$$

$$M_D = U_0 Q Q/ (4 \pi R_U 4 \pi) \qquad (18\text{-}36)$$

We see that the Mass of a dot is related to the permeability U_0, the charge Q to the second power and the inverse of the radius of the universe. From equation 18-36 it is clear that mass is a magnetic effect.

Section 18-6: The Mass of the Universe

The mass of the universe for the model universe can be calculated using Equation 18-20. Hence

$$M_U = (R_U C^2)/ G \qquad (18\text{-}37)$$

where M_U is the mass of the universe in Kg for the MKS system of units, R_U is the radius of the universe in meters, and G is the gravitational constant.

This equation is for a light sphere expanding from a pinpoint at the speed of light. It is a solution to an e^x curve.

We can solve for the mass of the model universe if we know the radius of the universe or the time of the universe from the pinpoint. The same is true for a universe, of e^{-x} a minimum to e^x a maximum since the time would be T_0 and normalized to the point where $e^x = e^0 = 1$. We can normalize our solution to the time equal to zero. In this case, everything before us is negative time leading to minus infinity while everything ahead of us is positive time leading to plus infinity. In this way we can study a universe which comes from minus infinity and goes to plus infinity. It is the exponential function e^x that permits easy normalization for the model of the universe.

We need to normalize the time of the universe to some quantity. We assume that we are at Time zero on the e^x curve.

A premise of the Dot theory is that coulombs decrease as the time increases. Thus the quantity coulomb second is a constant. This shows itself up in the slowing of all clocks in the universe and the slowing of the internal photon oscillation for the red shift while at the same time the ruler expands.

The red shift has two features that will increase the wavelength of the photon and one feature that will reduce the wavelength of the proton. We can say that the red shift is due to the expansion of the universe and it's Doppler effect. In that case the loss of photonic energy is negated by the increasing size of the ruler. Otherwise we can say that the increasing of the size of the ruler negates the Doppler Effect and the red shift is due to the loss of energy of the photon per unit distance. Both solutions produce the same results.

As the charge in the universe decreases the time increases. Therefore:

Coulomb Second = Constant   (18-38)

Although the math to derive the exact equation tends to be very complex we can write the solution by inspection. There must be a 4 pi term due to the spherical nature. We also most have a phase angle such as 30degrees because we are dealing with vectors. Therefore from the answer is obviously:

$$4 \pi Q T_U = \text{Cosine } 30 *(\text{coulomb sec}) \qquad (18\text{-}39)$$

In equation 18-39, Q is the standard charge of 1.60218E-19 coulombs and T_U is the time of the universe and Cosine 30 represents the vector phase angle. Using equation 18-39 we normalized the time of the universe to be 4.301396E17 or 13.63 billion light years.

Equation 18-39 is a very basic equation which shows a relationship between the charge Q and the time Tu. This solution is an overall solution. The actual solution involves the three universes. This causes periods where the light speed and charge are constant and then transformations in which the light speed increases in steps and the charge decreases in steps. In effect the universe is constantly reinventing itself.

Since the astronomers get 13.7 billion years from their linear analysis the cosine of 30 degrees appears to be the correct angle. Therefore we are dealing with a complex mathematical vector problem in which the vectors are 30 degrees apart.

The time of the universe then becomes:

$$T_U = T_O \, e^{(t/T_o)} = 4.30139E17 \text{ seconds} \qquad (18\text{-}40)$$

In equation 18-40 we see that if today t=0, the universe is T_O or 13.63 billion years old, and for a pure e^x solution this is always true. Corrections can always be made because the time (t) could be a million years or a billion years since the time when T_U equaled T_O. However this doesn't change how the universe works. It merely changes the exact point on the curve we are on. We may not have the absolute measurements but we know the general normalized answer anyway.

The radius of the universe is simply:

$$R_U = C \, T_U \qquad (18\text{-}41)$$

This calculates to be:

126

$$R_U = 1.28952E26 \text{ Meters} \qquad (18\text{-}42)$$

Using equation 18-37 and G=6.67260E-11, the normalized mass of the universe is:

$$M_U = 1.736896E53 \text{ kilograms.} \qquad (18\text{-}43)$$

We can now calculate the equivalent mass of a dot from equation 18-30.

$$M_D = K Q Q / 4\pi R_U C C \qquad (18\text{-}44)$$

In equation 18-44 we repeat the terms instead of using the square for clarity. Thus the Mass of a dot M_D is equal to the coulomb constant K times the charge Q squared divided by the speed of light C squared, divided by the radius of the Universe R_U, and finally divided by 4 π.

Using K=8.98756E9, Q=1.60218E-19, C=2.99792E8, R_U=1.28952E26, and π =3.14159, we get:

$$M_D = 1.58411E\text{-}72 \text{ Kg} \qquad (18\text{-}45)$$

The number of dots in the universe is the mass of the universe divided by the mass of each dot:

$$N_U = M_U / M_D \qquad (18\text{-}46)$$

$$N_U = 1.09628E125 \qquad (18\text{-}47)$$

In equation 18-47 we have broken the universe down into a large amount of dots of electromagnetic energy quanta by the normalization process. At this point in time the entire model universe has been quantized into tiny bits and pieces of energy and charge.

The number of dots in the neutron is the mass of the neutron, 1.67493E-27kg divided by the mass of a dot:

$$N_N = M_N / M_D \qquad (18\text{-}48)$$

$$N_N = 1.05731E45 \qquad (18\text{-}49)$$

There are a huge number of dots in one neutron. The number of neutrons in the universe can be calculated using the mass of the universe and dividing by the mass of the neutron. This is an equivalent number since a proton and electron plus some photon energy is considered a neutron.

$$\#N = M_U / M_N \tag{18-50}$$

$$\#N = 1.0370E80 \text{ neutrons per universe} \tag{18-51}$$

Section 18-7: The proton, the electron, and the photon

The number of dots in the neutron is:

Dots neutron = 1.057312E45   (18-52)

The corresponding number of dots for a particle is:

Dots particle=DotsNeutron x Mass Particle/ Mass Neutron
$$\tag{18-52}$$

Dots proton = 1.055854E45   (18-53)

Dots electron = 5.75037E41   (18-54)

where M_P = 1.67262E-27, M_E= 9.10939E-31, M_N = 1.67493E-27, and N_N = 1.22090E45.

If we know the mass of any particle we can find out how many dots there are within the particle. Likewise if we know the energy level of a photon we can find out how many dots make up the photon.

The electron has 5.75037E41 negative dots within it. The charge of each dot is:

$$Q_d = Q/ \#dots = 2.78622E-61 \text{ coulombs} \tag{18-55}$$

Equation 18-55 gives us the quantum of charge in the universe. This is the lowest amount of charge at the present time. Right after the big bang this quantity was very large since charge decreases with time.

The proton has the same amount of surplus positive dots as the electron. The proton has more total dots. Thus:

Total dots proton = 1.055854E45   (18-56)

Total excess positive dots = 5.75037E41   (18-57)

Difference = 1.055279 E45   (18-58)

Half difference = 5.276395E44   (18-59)

The proton contains the half difference of negative dots. The proton has the half difference and the excess positive dots. In general a photon has an equal number of positive and negative dots.

The only difference between matter and photons are the geometric configuration of the dots. As will be explained in more detail in Chapter 19 the property of mass is equal to a moving charge. As a dot moves around within an electron, it will form a three dimensional pattern. This can be looked on as an electrical gyroscope. The electron will resist outside forces moving it.

If we look at a collection of dots, the plus and minus dots can move in such a way that they nullify the resulting magnetic field. On the other hand if a plus dot moves in the opposite direction of a minus dot, the magnetic fields will increase instead of nullifying.

The plus and minus dots in matter aid and assist each other. The positive dots will flow in the same direction and the minus dots will flow in the opposite direction.

The photon can be viewed as dots moving in a plane perpendicular to the direction of motion. The plus and minus dots wipe out the resulting magnetic field but in so doing act like a self propelled drill bit that cuts through space as it moves ahead at the speed of light. Space itself is not empty. It is filled with a spectrum of dot-waves up to light sped infinity which causes impedance that resists electromagnetic waves from passing through it.

The Dot-wave theory is not concerned with subparticles. It is only concerned with the structure of the universe. All the particles and subparticles are made form dot-waves. Quantum mechanics theory has showed fairly well how the subparticles interact with each other and with the main particles.

Although the dot-waves are easy to understand, the various combinations of dot-waves and the dimensions they occupy produce all the amazing things that we see in the universe. Quantum mechanics is the means by which the interactions of the particles within our universe are explained. The measurements prove that Quantum mechanics give accurate results.

The dot-wave theory looks at our universe from the position of the mind of God. The source of the dot-waves is external to where they appear. Therefore the dot-wave equations are inverted. They show that bits and pieces of the mind of God produce particles and sub-particles which appear to us as our laws and properties of physics.

The result is that the dot-wave theory is a metaphysical understanding of the physics of the universe from a perspective of being outside the universe. It is an effort to understand how God created the universe out of the mind of God. All the dot-waves are little bits and pieces of the mind of God.

Although the calculations show that a proton contains a huge number of dots, what does that really mean? You will not find anything inside of particles other than what appears to exist in our experiments and measurements.

The total radius of our universe is twice the distance from where we are to the center of the universe. Each dot is the center of a sphere of Radius Ru. This means that there are a huge amount of spheres that make up the universe.

It is assumed that if we multiply the number of dots per neutron by the number of neutrons we will get the total number of dots in the universe. That may be true. It is also possible that many neutrons will exist within the center of some individual dot-waves and many other common dot-waves. This would reduce the total number of dots required in the universe.

Another way of looking at the dot-waves is to consider the universe to have a tiny mirror at the center and a huge surface mirror at a distance of 2Ru from the center. We live at a distance of Ru from the center. Photons within the universe will produce complex patterns as they move from the center to the outer reflective surface and back again to the center.

The dot-wave theory is an attempt to look at the source of our existence. It looks for the basic building block of the universe which is beyond our ability to measure and verify. To make things more complicated, String theory has made us understand that there are many different dimensions. What does that mean?

If we look at the universe as a sphere of total radius 2Ru, we can produce another dimension by having a separate sphere with a radius slightly greater than 2Ru. The distances between the spheres would be large distances apart but focal points upon the surface of our universe would only be tiny distances apart. The String theory mathematicians see many tiny distances but when we invert things the distances get large.

Our particles are then composed of focal points from many different simple universes which are electromagnetic fields. The interactions of the focal points as they move in different patterns make it appear to us that they are moving in tiny different dimensions. Since we only see the probability function of Quantum mechanics, a particle appears in its time and place.

From Heisenberg's Uncertainty principle we could not measure with great certainty both the position and the momentum of a particle. This caused what appeared to be a simple physical universe which obeyed the laws of classical physics to be looked upon in a strange manner.

The reason for this is that we look at the universe from what we are composed of. We try to be an independent observer and look at the universe objectively. However this is not easily accomplished because we are biased in a subjective manner due to our own physical limitations. It is not easy to look at a rock and say that there is nothing at all within the rock.

The dot-wave theory presents a different world view of the universe. It is an effort to understand the fundamental structure of the physical universe that accounts for the strange things that we see in the laboratory. This theory is just another way of trying to find the basis of our existence. The theory is not perfect. It is just an attempt to look at our universe from an inverted image as the mind of God would view the universe.

It is also an effort to look at the universe from equivalent models and from numerical analysis. Numerical analysis specifies that the universe is put together so that the constants of the universe are related to each other with simple formulas and simple constants. The gravitational constant is one example of where the constant is related to the speed of light, the radius of the universe, the mass of the universe and the simple constant of the square root of three divided by two.

The dot wave theory helps us to understand why it is so difficult to break a proton apart and attain its energy in the form of photons. There is really no difference between a proton and a photon. Both have the same type of plus and minus dot waves. The photon is very flexible energy whereas the proton is fixed energy.

If the proton only existed within the present universe then it would be pretty easy to tear it apart and obtain the tremendous amount of energy within it. The problem is that the three universe structure locks the proton into a sandwich of the past, the present, and the future. Instead of just a simple particle in the present, we get a structure of three protons tied together. This causes the proton to be bound into a structure that was born at the big bang or right after the big bang.

At CERN we use tremendous energy to break apart and produce matter and antimatter. Yet they are all the same things. Some things we produce do not live very long. The most likely solution is that these particles are the product of photons which only live in the present.

Photons are very soft energy. They become part of particles and then leave them. The photons were always free. The compressive forces at the big bang locked the proton into place. It may be possible that some machine could free the proton and release all the dot-waves in photonic form. Perhaps it is only a dream. Perhaps this only happens upon higher Earths where there is less mass and higher photonic energy.

The dot-wave theory provides us with the answer to gravity. It does not seem to have much use in helping us toe understand the various particles and subparticles. Quantum mechanics appears to give excellent answers in that regard.

Gravity is just a simple electrical engineering problem. Perhaps that is why the greatest scientific minds have found gravity so difficult to understand. It is just a simple problem that you just write the answer for. The detailed math of spherical vector fields can be quite complex from a mathematical perspective but from an Engineering perspective we only need the equations and an understanding of how gravity works. It is not really necessary to use very fancy math when simple algebra solves the problem. Perhaps that is why Einstein and other great mathematical physicists had so much difficulty with the gravity problem.

Even today people are searching for a particle which produces mass and gravity such as the graviton. No such particle is necessary nor will such an entity be found. The Bohr orbit has all the parts necessary to solve the gravity problem. Once we add Hubble to Bohr the problem is solved. This will be shown in the next chapter as we convert mass in kilograms into charge velocity in coulomb meters per second.

CHAPTER 19 – CONVERSION OF MASS TO CHARGE

SECTION 19-0: INTRODUCTION

In this chapter simple Engineering type equations for the Dot theory will be used to find the conversion from mass to charge. The universe is an electrical universe and the units of kilograms can be described electrically. At present the universe is specified in terms of kilograms, coulombs, meters, and seconds. There are many different names for things such as Amperes but Amperes are really coulombs per second. In the usual mechanical/electrical physics books, four units are required. Once we convert mass in terms of coulombs, meters, and seconds then only three units are required.

The mass to charge conversion is limited to the constraints of the universe that we live in. If the universe is expanding with increasing ruler and time clock while at basically constant light speed, then we have fewer conversion choices. In Section 19-1 Dot Theory conversion equations are used. In Section 19-2 some general mass to charge conversions are looked at. They represent a whole class of conversions but only those with a basically constant light speed solution have been studied in detail. Although the universe jumps to higher and higher light speeds as it recreates itself, at every present universe it looks like a constant light speed universe.

Section 19-1: Conversion Theory

Most of the equations in Chapter 18 are a set of normalized equations for a model universe that expands at a basically constant light speed and with an expanding ruler and an expanding time clock. Some of the equations such as equation 18-34 are standard electrical equations, whereas others are for the normalized model universe. These equations serve to reduce the equations of physics into three variables of kilograms, meters, and seconds; or coulombs, meters, and seconds. Using these equations, it will no longer be necessary to use combined electrical and mechanical units. Although this is important to study to understand the basis of the universe, the mechanical equations are much easier to use for ordinary mechanical problems.

In order to eliminate either kilograms or coulombs, it is necessary to produce a table of conversion. We can replace kilograms with some function of coulombs, meters, and seconds. Thus we can define mass in terms of charge and some power of light speed. Charts can then be produced using many combinations that match the equations in Chapter 18 as far as units are concerned.

The process for the production of the charts is readily accomplished. The comparison study of the charts takes many years to accomplish. Each chart will produce equations for the universe that may or may not represent reality. Fortunately the charts tend to be dual or sister solutions so things learned from an incorrect chart still applics to our universe. Thus all the solutions can be considered Sister solutions. However, only one solution is the actual solution and this is called the Sister 1 solution.

As a Sister 1 solution we could say that mass has the units of coulomb meters/second and that energy has the units of coulomb meters^3/seconds^3. This solution meets the Dot theory criteria of basically constant light speed since as meters increase and seconds increase, the light speed remains constant. The mass then only depends upon the coulombs, which decrease as the universe expands. This solution says that mass is a property of moving charge. It also says that energy is charge moving in a volumetric fashion.

For a Sister 2 solution, a dual solution could be looked at. Mass would be coulomb seconds/ meter and energy would be coulomb meters/ second. In this solution charge would be a property of a moving mass.

In general for a Dot theory solution, mass must equal charge times any power of light speed. Therefore,

$$M_D = Q_D C^n \qquad (19\text{-}1)$$

where C^n could be any positive or negative power of the light speed. The corresponding dot energy is:

$$E_D = Q_D C^{n+2} \qquad (19\text{-}2)$$

In equation 19-2 we see that once the form of the dot mass is chosen, the energy is C^2 higher since $E=MC^2$.

We can now produce a chart of the Sister 1 solution and the Sister 2 solution to see the various relationships. In general the Sister 1 solution states that:

$$M = Q C \qquad (19-3)$$

Equation 19-3 states that the unit of mass is charge times the speed of light. This basically states that a moving charge causes the property of mass. The dual solution or Sister 2 solution is:

$$Q = MC \qquad (19-4)$$

Equation 19-4 states that the unit of charge is mass multiplied by the speed of light. This basically states that a moving mass causes the property of charge.

I have investigated many other solutions over the years but the simplicity of equations 19-3 and 19-4 causes either one to be considered a most likely solution. However, the Sister 1 equations will be shown to match the physical universe more readily.

Charts of other solutions are easily prepared. However, once higher powers or square roots or cubes of the light speed occur, they lack the simplicity of the Sister 1 solution or the dual Sister 2 solution.

The important thing in the method shown is that the equations presented so far are unit's equations in which only the three ingredients of kilograms, meters, and seconds, or coulombs, meters and seconds are used. Since the mechanical world interlocks with the electrical world at energy and force equations, it was never necessary to have so many different units. At most we only needed kilograms, coulombs, meters, and seconds. By specifying that the universe is completely electrical, only three units are needed.

Let us now look at a table of the two dual most likely solutions for the relationship between mass and charge.

The following table of units is made for the most likely Sister 1 solution with Mass = Charge times Velocity and the dual Sister 2 solution with Charge = Mass times Velocity.

Table 19-1: Conversion of Mass to Charge (MCS system))

Quantity	Sister 1	Sister 2
Mass (M)	Cou Met/Sec	Cou Sec/Met
Charge (Q)	Coulombs	Coulombs
Energy (E)	Cou Met^3/ Sec^3	Cou Met/Sec
Coulomb Const.	Met^4/ Cou Sec^3	Met^2/CouSec
Force (F)	Cou Met^2 / Sec^3	Cou /Sec
Momentum (MV)	Cou Met^2 / Sec^2	Coulombs
Plank's Const (h)	Cou Met^3 / Sec^2	Cou Met
Permeability (U$_O$)	Met^2 / Cou Sec	Sec/Cou
Permittivity (ε_O)	Cou Sec^3 / Met^4	CouSec/Met^2
Voltage (V)	Met^3 / Sec^3	Met/Sec
Current (I)	Cou / Sec	Cou/sec
Impedance (Z)	Met^3 / Cou Sec^2	Met/Cou
Grav Const (G)	Met^2 / Cou Sec	Met^4/CouSec^3
Power (P)	Cou Met^3 / Sec^4	Cou Met/Sec^2
Flux Density (B)	Met / Sec^2	1/met
Inductance (L)	Met^3 / Cou Sec	Met Sec/Cou
Charge/Mass (Q/M)	Sec/ Met	Met/Sec
Capacitance ($\underline{C}$)	Cou Sec^3/ Met^3	Cou Sec/Met

In the table the various quantities have been shown in the Meters Coulombs Seconds (MCS) system for the most likely Sister 1 and Sister 2 solutions.

Section 19-2: General Mass to Charge Conversions

In Section 19-1, the Dot theory method of matching mass to charge was formulated. The basis of the method was that charge and mass both decreased with time or distance as the universe expanded. The relationship between mass and charge then became ratios of the speed of light to various positive and negative powers.

A more complete solution of various possibilities would be that mass also could vary with charge or current and a power of the speed of light. Thus more possibilities exist beyond the Dot theory although they become less probable. Let us look at the simplest general possibilities. The following are possible:

Mass = Charge   (M=Q)   (19-05)

Mass = Current = Charge per second(M=I)   (19-06)

Mass = Charge per meter   (MC=I)   (19-07)

Mass = Current x Velocity   (M=IC)   (19-08)

Mass = Charge /Velocity   (MC=Q)   (19-09)

Mass = Charge x Velocity   (M=QC)   (19-10)

We can now make charts of all these conversions by using the following standard equations:

$$Force = KQQ/R^2 \qquad (19\text{-}11)$$

$$h = Energy \times Time \qquad (19\text{-}12)$$

$$GMM = KQQ \qquad (19\text{-}13)$$

$$V = KQ/R \qquad (19\text{-}14)$$

$$U_0 \varepsilon_0 = 1/ C^2 \qquad (19\text{-}15)$$

$$B = U_0 \, I / \, R \qquad\qquad\qquad (19\text{-}16)$$

Equations 19-11 through 19-16 are standard physics equations and enable us to produce charts of the various relationships in terms of meters, coulombs, and seconds for the MCS system. We can also chart the modified GG/MKS system of meters, kilograms, and seconds.

The solution for mass = charge, (Equation 19-05), tells us that a moving mass produces currents, and positive and negative magnetic fields from positive and negative dots. Since mass is loaded with dots, one characteristic of mass is identical to charge itself. However, when we look inside the proton, we see dots in constant motion. We see current flows and they look like current gyroscopes. Thus a primary characteristic of mass is not charge but it is related to the motion of charge. Although (mass = charge) falls within the Dot theory since both mass and charge decrease with an increasing universe, it is not a real possibility.

The solution mass = current, (Equation 19-06) is a characteristic of mass. If we look inside the proton, we do find currents within it. Yet, what give them the gyroscopic nature are not currents themselves but currents flowing in a circular path. This solution is not a Dot theory possibility since mass and charge do not track each other as the universe expands.

The next possibility for the primary electrical characteristic of mass is that momentum is current, (Equation 19-07). Moving masses do have moving dots and currents exist. That mass is a coulomb per meter can be considered a characteristic of mass. However, we are looking for a gyroscopic effect for the electrical equivalent of mass. This solution is not a Dot theory solution but it is interesting to study and compare to the other solutions.

The next possibility is that mass is current times velocity, (Equation 19-08). In this case, as the universe expands the mass drops and the charge remain constant.

This solution has all independent units. No basic relationship between the gravitational constant and the magnetic field is self-evident. It does provide an alternate solution; however is not part of the Dot theory.

The next possibility is that mass equals charge over velocity, (Equation 19-09), or that a moving mass creates charge as a dual Sister 2 solution. It provides a good electrical to mechanical analogy with mass and capacitance being equal. It remains the secondary dual solution for comparison study and analogy since the differential equations of capacitive circuits are of identical form as those equations for mass.

Let us now look at the last solution, (Equation 19-10) which is the Sister 1 solution. In this solution a moving charge produces mass. Thus:

$$\text{Charge x Velocity} = \text{Mass} \qquad (19\text{-}17)$$

Equation 19-17 is for charge momentum. It is the electrical dual of a physical momentum. Thus:

$$\text{Mass Momentum} = \text{Mass x Velocity} \qquad (19\text{-}18)$$
$$\text{Charge Momentum} = \text{Charge x Velocity} \qquad (19\text{-}19)$$

We see in equations 19-18 & 19 that both mass and charge momentum's are similar quantities. Breaking down equation 19-19 into parts, we get:

$$[\text{Charge/second}] \text{ x Meters} = \text{I x R} = \text{Mass} \qquad (19\text{-}20)$$

In equation 19-20 we see that we have a current (I) operating at a radius R that gives us a current torque or charge momentum. This is a strong possibility for mass since we can see within the proton, currents flowing in circular paths and oscillating from inner radius to outer radius as well. We also see a root mean square current at a particular radius. Thus the current gyroscopic action within the proton adds credibility to equation 19-20 as being a primary conversion equation. Let us now produce a chart of the various relationships for this Sister 1 solution for both the electrical (MCS system) and the mechanical (GG-MKS system). In the later, the MKS system has been modified to show charge (Q) in terms of mechanical units where charge is kilogram seconds per meter.

TABLE 19-2: MASS TO CHARGE CONVERSION (M=QC)

Quantity	MCS-System	GG-MKS-System
Mass (M)	cou met/sec	kg
Charge (Q)	coulombs	kg sec/ mct
Charge/Mass	sec/met	sec/met
Velocity	met/sec	met/sec
Acceleration	met/sec^2	met/sec^2
Energy (E)	cou met^3/sec^3	kg met^2/sec^2
Force (F)	cou met^2/sec^3	kg met/ sec^2
Momentum	cou met^2/sec^2	kg met/sec
Plank's (h)	cou met^3/sec^2	kg met^2/sec
Coulomb (K)	met^4/cou sec^3	met^5/kg sec^4
Permittivity	$\varepsilon_0$cou sec^3/met^4	kg sec^4 / met5
Permeability	met^2/cou sec	met^3/ kg sec^2
Grav. Const	met^2/cou sec	met^3/kg sec^2
Voltage	met^3/sec^3	met^3 / sec^3
Current	cou/sec	kg/met
Impedance	met^3/cou sec^2	met^4/ kg sec^3
Inductance	met^3/cou sec	met^4 / kg sec^2
Capacitance	cou sec^3/ met^3	kg sec^4/ met^4
Flux Density	met/sec^2	met/ sec^2

In Table 19-2 the gravitational constant has the same units as electrical permeability. This shows that gravity is a magnetic force. In addition flux density (B) has the same units as acceleration. This shows that an accelerating space-time electromagnetic field produces magnetic flux. Since voltage is cubic velocity, this shows that the moving universe generates voltage. Thus the motion of the electromagnetic field produces currents and voltages and magnetic flux as it expands.

The relationship between the gravitational constant and the electrical permeability using a best numeric fit is:

$$G = 16 \pi e U_O / (137.036)^3 = 6.6720E\text{-}11 \quad (19\text{-}21)$$

Equation 19-21 shows the exact relationship between the gravitational constant and the electrical permeability constant. It shows that the gravitational field is purely magnetic and that the conversion chart in Table 19-2 provides us with a very important conversion from mass to charge.

In the Sister 1 (MCS) solution mass per charge as per Equation 19-17, is a constant as the universe stretches out with both time and distance expanding. Energy is volumetric in both time and distance and drops as coulombs drop. Thus:

$$MV^2 = Q V^3 \quad\quad\quad (19\text{-}22)$$

In Equation 19-22, energy has the units of charge times velocity cubed. Likewise it is charge within the confines of cubic meters over cubic seconds. Thus energy is a volumetric space-time entity. It is charge oscillating in three dimensional distance (meters) volume over three dimensional time (seconds) volume.

$$MV = Q V^2 \quad\quad\quad (19\text{-}23)$$

In Equation 19-23, for momentum, coulomb meters squared per seconds squared would be charge oscillating in two dimensional distance (meters) area over two dimensional time (second's) areas. Thus all the various units mean something with respect to dimensions of velocity which will be explained in section 19-3.

The Sister 2 solution has the simplest units. However, the more complex units of the Sister 1 solution provide us with a greater understanding of space and time. Since there are three dimensions of light speed as will be explained later, there are three dimensions of distance and three dimensions of time also. This helps to explain the triple universe.

The fact that energy has the cubic units for the Sister 1 solution makes that solution more important. The Sister 2 solution is a great duality and since $M_N C = \pi Q$ approximately, it appeared most important originally. The Sister 1 solution matches at the gravitational constant / electrical permeability constant. However, that is what was sought since this is where the mechanical world matches the electrical world. Of course we must find what mass matches what charge.

The sister 1 solution enables us to see that mass is the little current gyroscope within the proton and electron. In the photon, you may get circular action perpendicular to the plane of motion but no circular action in the front to back region. Thus the photon will spiral and move forward and have no mass in the front to back direction. The photon will merely be a planar current loop which possesses no three dimensional gyroscopic ability.

The five solutions investigated all are part of the general conversion equation from mass to charge. Thus:

$$GMM/R^2 = U_O \, QV \, QV/ \, R^2 \qquad (19\text{-}24)$$

Equation 19-24 is a general unit's equation of standard physics that relates the gravitational force to the electrical magnetic attraction. The units are standard physics. The five most probable solutions to this equation are:

$G = U_O$	($M = QC$)	(19-25)
$G = U_O R$	($M = QC/ \, R$)	(19-26)
$G = U_O C^2$	($M = Q$)	(19-27)
$G = U_O C^4$	($MC = Q$)	(19-28)

$$G = U_O R^2 C^2 \qquad (MC = QC/R) \qquad (19\text{-}29)$$

If G and U_O are reasonably true constants, they should not be a function of the distance R with respect to each other. Thus equations 19-26 and 19-29 are less likely to have much meaning. In addition, the distance R that would match G and U_O is not of the order of the Bohr orbit or the proton radius. When we look at equation 19-28, we see that G would be of the same form as the electrical permittivity. In addition, if both M and Q were identical, we would be at a loss to explain all the properties of the universe.

The net result is that the Sister 1 solution of Equation 19-25 and the dual Sister 2 of equation 19-28 form the possible primary characteristics of the mass to charge conversion. Since the numeric relationship between G and U_O in equation 19-21 is merely the 16 π e term and the 137.036 term, it appears as the most important solution. Of course this does not deny the importance of the dual solution of equation 19-28 since the masses are inter-related by C^2.

For the rest of the book the Sister 1 solution will be used as the primary solution which relates charge to mass and the constants of the universe. We can then calculate the constants of the universe in terms of each other. Thus repeating equation 19-21:

$$G = 16 \pi e\, U_O / (137.036)^3 \qquad met^2/cou\ sec \quad (19\text{-}30)$$

Equation 19-30 is the best fit for the numerical conversion from the gravitational constant to the electrical permeability constant. This calculates to be:

$$G = 6.67223E\text{-}11 \qquad\qquad (19\text{-}31)$$

In standard physics, the relationship between Z_o and h and Q is:

$$Z_O = 2h/\ 137.036\ Q^2\ = 376.828\ met^3/cou\ sec^2 \quad (19\text{-}32)$$

From the fine constant relationship in standard physics, we get:

$$hC/\ [2 \pi K Q Q] = 137.036 \qquad\qquad (19\text{-}33)$$

The 137.036 comes from the reciprocal of the cosine of 360 degrees over 274 half waves as was discussed in Chapter 17 equations 17-1 through 17-4. Thus:

$$137/Cos\ (360/274)=137/0.999737 =137.036028 \quad (19\text{-}34)$$

The relationship between G and the electrical permittivity constant becomes:

$$G = 16\ \pi\ e\ /\ [\varepsilon_0\ C^2\ (137.036)^3\ ] \quad (19\text{-}35)$$

This calculates to be:

$$G = 6.67224E\text{-}11 \quad (19\text{-}36)$$

We can also relate Z_0 and GC, thus:

$$Z_0 = [GC\ (137.036)^3]\ /\ 16\ \pi\ e \quad (19\text{-}37)$$

GC calculates to be:

$$GC = 2.000392E\text{-}2\ = 1/\ 50\ ohms \quad (19\text{-}38)$$

$$Z_0 = 376.751\ met^3/cou\ sec^2\ (ohms) \quad (19\text{-}39)$$

We see that the term GC is an admittance of approximately $1/50$ ohms. We can also write an equation for h in terms of GC and Q^2.

$$h = GCQ^2\ [(137.036)^4]\ /\ 32\ \pi\ e \quad (19\text{-}40)$$
This calculates to be

$$h = 6.62647E\text{-}34 \quad (19\text{-}41)$$

We can also add our standard electrical equation:

$$\varepsilon_0\ U_0 = 1/\ C^2 \quad (19\text{-}42)$$

By feeding the equations into each other using the standard formulas of physics, we can derive all the above equations from the Sister 1 table of units. The net result is that the Sister 1 solution enables us to have a whole set of interlocking equations that relate h, c, Q, Uo, ε_0, Zo, and G. The Sister 1 solution enables all the main constants of the universe to be inter-related. This was not accomplished using the Sister 2 solution or any other solution investigated. This leads one to believe that the Sister 1 solution is the correct solution for the present universe.

It is important to understand that the mind of God has many different choices in the selection of the manner by which the dot-waves operate. Each sister solution will produce a different type of universe. In addition the basic dot-wave equations can change as well. The dot-wave theory gives us a little insight into the inner workings of God's mind. As God thinks of a physical solution the dot-waves will form patterns to produce what we see and measure.

Section 19-3: The Lightspeed Equations

For the Sister 1 solution we have units of meters cubed over seconds cubed and meters squared over seconds squared and various combinations of powers of meters and seconds. What do these units mean? In this section we will investigate the meaning of such complex units.

When the radius of the universe increases, both the charge and the energy of the universe drop. Looking back in time, we see that as we compress space-time, we charge up the universe and give it energy. We can look at energy from equation 19-22 as charges moving in a space time volume. Thus:

$$\text{Energy (E)} = Q\ C^3 \qquad (19\text{-}43)$$

In equation 19-43 we see that energy is charge times the speed of light cubed. In addition:

$$E = M\ C^2 \qquad (19\text{-}44)$$

In Equation 19-44 we have Einstein' famous equation and energy is mass times velocity squared. Also from equation 19-43 we see that energy is charge time velocity cubed. Thus solving for M we get:

146

$$M = QC \qquad (19\text{-}45)$$

Equations 19-43 and 19-45 are the missing piece of the puzzle that Einstein started to solve. For momentum the equation is:

$$MV = QVC \qquad (19\text{-}46)$$

Finally for the momentum of a photon, the equation is:

$$M_O C = Q C^2 \qquad (19\text{-}47)$$

In equations 19-45 we find that mass in coulomb meters per second is a first order linear function of the light speed. In equation 19-47 we see that momentum in coulomb meters squared per seconds squared is a planar function of light speed. Finally in equation 19-43 energy in coulomb meters cubed per seconds cubed is a volumetric function of light speed. These equations permit the reader to see how the entire spectrum of possible universes varies with light speed.

For constant energy solutions, from equation 19-43 we see that charge decreases with the cubic power of light speed. If the operating point of this universe was at 2C or twice our present light speed, the charge of the electron would be 1/8 that of our present charge.

The above equations are light speed C equations. They tell us the properties of the coexisting universes for the triple universe solution. They also tell us what the universe will become as the universe moves up toward infinity light speed. They also help us to understand the condition of the universe right after the big bang when the universe was pure mass exploding into energy. They also help us to understand the conditions at the little bang when the universe loses all mass and become pure energy in chaos.

The above equations are the primary equations of the universe. Einstein was able to present one primary equation (Equation 19-44). This book presents the rest.

Section 19-4: the Gravitational Constant

We can now show how the equation for the gravitational constant was derived using electrical theory and the mass to charge conversion charts in section 19-2. This gives us

147

an exact equation that defines the exact relationship between the electrical world and the mechanical world.

We can produce a general electrical equation from magnetic field attraction theory. The exact constant such as π, 4π, 0.866, depends upon the geometry involved. It is necessary to find the best electrical fit that produces a universe in the order of 13.7 billion years as per the astronomical measurements. The Hubble telescope has improved the accuracy of the time estimate greatly since when I started this effort in 1981. This method will provide us with a second means of deriving the time of the universe since big bang.

Years of studying the constants of the universe by numerical analysis and a simple engineering calculator has demonstrated to me that the universe is tied together by the constants 0.866, 1.414, e, π, 4, 16, 137, and 137.036. These numbers always appear in various formulas.

Numerical analysis shows us that the above factors enter into many equations of modern physics. I accept such constants as being appropriate.

In general the gravitational force is driven by the discharge of the dot charges throughout the universe. As the universe expands dot current flow of both positive and negative nature extends from all matter and photons outward to the radius of the universe. This causes spherical magnetic fields that are constantly expanding. The space-time memory of the past magnetic field, the present magnetic field, and the future magnetic field within the three universes produces a uniform force. These forces can be looked at from an electrical perspective or a mechanical perspective. They are two different ways of looking at the same thing.

This force can also be viewed as the contracting force acting upon the Bohr Orbit when we view the universe from the mechanical perspective. Therefore we can write the gravitational field equations in terms of the expansion of the Bohr orbit.

At the same time as the universe expands, the Bohr orbit expands and this causes a force between the present orbit, the past complete orbit, and the future orbit. Likewise a similar force exists between the past proton inner oscillation, the present inner oscillation, and the future inner oscillation. They can all be looked at as Ampere's laws current loop problems. The force between past, the present and the future tends to both expand the proton and contain the proton from flying apart. These forces appear to be binding energy type forces.

The hydrogen atom becomes the standard because there is a simple coulomb attraction force and gravitational forces as well. Since the Bohr atom is expanding, the protons are also expanding and the neutron radius is expanding as well. The general expansion of the universe common mode produces the decay of charge all over the universe and the general gravitational field. This can be viewed as the loss of mass within the particles or the loss of charge within the particles.

The gravitational force of concern is the force between two hydrogen atoms. This is the same force that operates upon two heavy metal balls in a lab. Likewise it is the same force which holds us to the Earth.

Let us look at the force between two hydrogen atoms by producing a magnetic attraction equation that relates the spinning electron of one atom interacting with the Bohr Orbit expansion of the second atom. One term will be large, the other small as shown in the following equation (19-48):

$$GM_HM_H / R^2 = 2U_O (QC/137.036) [4\pi \ QV_B{}^*] \ Cos30° / (R^2)$$

Equation 19-48 is a general electrical equation relating a gravitational force between two hydrogen atoms to the interaction of the spinning electron in the first Bohr Orbit of one atom producing one magnetic field represented by the term $QC/137.036$, and a magnetic current loop caused by the slow expansion of the Bohr radius of the second atom represented by $4\pi \ QV_B{}^*$.

The factor (2) is caused by the electron of atom 1 reacting with the field of atom 2 and the electron of atom 2 reacting with the field of atom 1. This doubles the force.

Since vectors are involved the typical electrical vector angle of 30 degrees is used. The velocity $V_B{}^*$ represents the motion of the Bohr radius as it expands slowly in time.

The constant G represents the gravitational constant of 6.67260E-11, M_H represents the mass of the hydrogen atom of 1.67353E-27, Uo is the electrical permeability of 1.25664E-6, C is the speed of light of 2.99792E8, Q is the charge of both the electron and the proton of 1.60218E-19, and R is the distance between the two atoms.

The equation for the electrical form of the gravitational force can be interpreted in two different ways. If we know the gravitational constant then we can obtain the time of the universe since big bang from the expansion velocity V_B^*. Likewise if we know the expansion time of the Universe from astronomical measurements we can obtain the constant G by working backwards.

From equation 19-49, we can solve for the velocity of expansion V_B^* of the Bohr orbit.

$$V_B^* = 1.21667E-28 \text{ meters per second} \tag{19-49}$$

The standard physics equation for the Bohr radius is:

$$R_{Bohr} = 137.036 \text{ h} / (2\pi M_E C) \tag{19-53}$$

The Bohr radius calculates to be 5.29178E-11 based upon an electron mass of 9.10939E-31 for the Bohr orbit.

The time of the Universe is:

$$Tu = R(Bohr) / V_B^* = 4.34940E17 \text{ seconds} \tag{19-50}$$

$$Tu = 13.7827 \text{ billion years.}$$

This method is within 0.58 percent of the astronomical data.

In the Dot theory we normalized the time of the universe by equation 18-39:

$$4\pi Q T_U = \text{Cosine 30 (coulomb seconds)} \tag{19-51}$$

This gives a time of:

$$T_U = 4.30139E17 \text{sec} \tag{19-51}$$

$$T_U = 13.6306 \text{ billion years} \tag{19-52}$$

The 13.6306 billion years matches the astronomical data of 13.7 billion years to 0.51 percent. The problem is that there is an assumption that a single big bang occurred and this continued to this day. In reality there was an initial big bang and then a series of cosmic events caused by the explosion of black holes in galaxies. At the same time the initial light speed at the first big bang was near zero. Therefore a much more complex calculation is required to produce the actually time since the first big bang. It may very well turn out that the answer approaches minus infinity. However if the system operated along an exponential curve, the best we could ever do would be a straight line approximation. Then it would be very difficult to get another answer unless we could live a very long time to see any differences.

In any event it is a good yardstick for the Dot theory. Any other means of measuring the time of the universe must match reasonable well. In equation 18-40 we assumed that time was following an exponential curve and that today was the zero time reference point. Thus we got a normalized time of 13.63 billion years.

If we assume that both methods have a slight error, we can take the geometric mean to arrive at a closer solution. The result is:

$$Tu = 13.7064 \text{ billion years} \qquad (19.56)$$

The difference between the geometric mean of the Dot theory normalized time and the Bohr atom analysis time is within an accuracy of 0.047 percent of the Hubble astronomical data of 13.7 billion years.

If we write equation 19-36 in terms of Uo, we get:

$$G = 16 \text{ pi e Uo}/ (137.036)^3 = 6.67223E\text{-}11 \qquad (19\text{-}57)$$

This expression gives G (6.67260E-11) to an accuracy of 0.005 percent. Here G is clearly shown as the electrical permeability times the constant:

$$\text{Constant} = 16 \text{ pi e} / (137.036)^3 \qquad (19\text{-}58)$$

This constant appears to be a conversion between a wave (16 pi e) and a Bohr radius moving in three different rotations for the X, Y, and Z axis. The exact mathematical reasoning is beyond my engineering abilities. However this type of number has always appeared during my numerical analysis of the proton and the neutron.

When we analysis the hydrogen atom, the Einsteinian mass increase for the electron at C/ 137 is:

Energy = 13.58 EV   (19.59)

For this case the binding energy of the electron is identical with the Einsteinian mass increase at the Bohr radius. This radiated away and the electron was bound. You then have to add back this energy to release the electron.

When we bring an electron to the surface of the neutron, we get an Einsteinian energy of:

Energy = 0.782 MEV   (19-60)

This number is equal to the mass of the neutron minus the mass of the proton and then minus the mass of the electron.

This shows that the energy of the neutrino is actually the Einsteinian energy of an electron at the neutron's radius. In effect this is the electron's speed of 0.9186 C at a radius equal to the neutrons wavelength of 1.319598E-15. This is why the neutron is unstable when isolated. It radiates out the Einsteinian energy which we call the neutrino.

The accuracy of the time of the universe by my hydrogen atom equation will vary due to differences between the hydrogen atom, the neutron, and complex structures of molecules in the universe including black holes and neutron stars... In addition there is an error band since no two platforms in the universe are identical. Therefore it appears to me that either of my two methods for the calculation of the time of the universe is within the allowable error band.

Chapter 20 Doppler Space Time

Section 20-0 Introduction

In this chapter we will look at the original work of Lorentz/Einstein and add to it Doppler Radar principles together with the classical linear space-time interpretation of the Michelson/Morley experiment. Einstein produced an orbital space-time interpretation of the Michelson/Morley experiment. This is a steady state interpretation involving cyclical motion. Basically Einstein's solution is a steady state electrical circuit solution. It is excellent for orbital motion and clocks moving around the Earth. It is not good for linear transient motion, which is defined by classical physics. This chapter involves ordinary physics and simple equations.

Linear space-time is important for future high-speed space travel. It is linear classical physics that determines the amount of energy required for traveling to nearby stars. Einsteinian orbital space-time works well in describing the Bohr orbit, the planetary motions, the clock in motion around the Earth, etc. It is not applicable for linear motion moving toward the speed of light. It limits man's ability to understand and achieve high-speed space travel.

To travel the stars is a dream of many. When we go up to higher Earths it is readily accomplished. Upon this Earth it is only a possibility since the physics upon the higher Earths permits particles and sub-particles to be readily used for fuel. Upon this Earth such things are difficult. In addition the very long lives upon higher Earths permit people to sleep while traveling huge distances. Our bodies cannot survive any long space fight to viable planets upon other stars. However this study is interesting and therefore worthwhile to help us understand the future possibilities here and elsewhere.

153

In this chapter, we will look at the space-time solution from a general single frequency describing function method in which the Y and Z dimensions are invariant. We will not include non-linear analysis, Fourier series, or spectrum analysis at this level of the work.

Section 20-1: Doppler Space Time Principles

The moving electromagnetic fields of an object produce inertia, kinetic energy, and all the other properties of the mechanical world. GG/MM/ Doppler space-time shows what the mass of an object looks like in the forward direction and also in the reverse direction. It is a complete space-time solution from an engineering viewpoint as opposed to Einsteinian space-time, which is a reasonably valid but partial solution.

The big problem with the Einsteinian space-time solution is that it applies steady state electrical theory as the general solution to space and time while ignoring transient electrical theory. In general the steady state solution is a special case of the transient solution. Einsteinian space-time would lead one to believe that the experiments in the cyclotron for orbital motion are valid for linear space travel. He ignored standard classical physics and applied orbital or steady state theory to linear motion. This produced the clock paradox and other misunderstandings of the true nature of space and time.

We know that light is mass-less in the forward or rearward direction. We know that light bends around stars and thus has mass in the perpendicular direction. Einsteinian space-time does not take into account the differences in mass in the X, Y, and Z directions. GG/MM/Doppler Space time does.

Let us look at a perfectly stationary sphere in pure outer space that has a gravitational field extending evenly in all directions. We can look at the sum total of all the inner fields and motions of all the electrons, protons, and

154

neutrons as equivalent to a single AC gravitational field located at the center of mass of the object. The minute we move this sphere, the gravitational magnetic field lines in front of the sphere become compressed while the gravitational magnetic field lines in back of the sphere become elongated. Of course every Bohr orbit and every proton, electron, and neutron experiences this effect

As you look at the Bohr Orbits, the front to back oscillation in the direction of motion starts to turn perpendicular to the direction of motion. The size of the X direction decreases while the size of the Y and Z direction increases. As more photons are added and the velocity increases, the Bohr Orbits and the individual neutron, electron, and proton oscillations flatten in the X direction of motion. As we move toward higher speeds, the X direction shrinks rapidly. At very high speeds, a sphere becomes a flat penny.

In this chapter we will consider only the shrinkage of the X direction. The Y and Z directions can remain the same size, they can shrink, and they can expand. All this will depend upon various constraints. In general if you compress an ordinary object in the X direction, the Y and Z dimensions will expand and the volume of material will attempt to remain constant. In this chapter we will look at the constant Y/constant Z solution for variable X. This will match the Einsteinian solution and it will enable the reader to see the basic error of Einstein in his space-time formulas.

If we keep adding photons to the ball, the front gravitational magnetic field lines will be crushed together while the back gravitational magnetic field lines will form a tail. The result will be a Fourier series complex gravitational field. An ordinary block of iron will form a complex spectrum of Fourier series gravitational fields. This is an extremely complex electromagnetic problem. However, complex things are often treated with simplified describing functions or equations, which reasonably explain the phenomenon.

Let us now write equations for the spherical object (ball of mass) which stands stationary. Let us assume we are deep in space away from any strong gravitational fields and clear of any rotation from the galaxies since all these produce errors. The ball can be represented as a total mass at a particular wavelength.

$$\lambda = C/f = h/MC \qquad (20\text{-}1)$$

In equation 20-1, the wavelength of the object at rest is equal to the speed of light divided by its frequency. This is also equal to Plank's constant h divided by the mass of the object and the speed of light. For example the wavelength of a neutron is 1.31959E-15. The corresponding frequency of the neutron is:

$$f_N = C/\lambda_N = M_N C^2 / h \qquad (20\text{-}2)$$

The frequency of the neutron is the mass of the neutron times the speed of light squared divided by Plank's constant h.

$$f_N = 2.27186E23 \qquad (20\text{-}3)$$

where M_N=1.67493E-27, C=2.99792E8, and h=6.62608E-34. The frequency of the neutron is very high. From Equation 20-2 we see that the equivalent frequency of a large ball would be huge. However, this is not true. The frequency of a large ball is the same as the neutron but the amplitude of the gravitational field is increased. The gravitational field of a huge ball is equal to the equivalent number of neutrons times the field strength of one neutron. Thus:

$$\text{Ball's Field} = N_N \times \text{Neutron Field} \qquad (20\text{-}4)$$

Let us now move the ball with a velocity V_B. If we place an observer in back of the ball and another in front of the ball, and measure the mass frequency of the ball, we get for a ball moving toward the right:

$$f_{ML} = [C/(C+V_B)] f_O \qquad (20\text{-}5)$$

$$f_{MR} = [C/C\text{-}V_B] fo \qquad (20\text{-}6)$$

In equations 20-5 and 20-6 we see that the Doppler Mass frequency of the ball is smaller to the left behind the ball and larger to the right in front of the ball. Although Einstein had a simple singular total mass, the reality is that a simple Einsteinian type mass does occur in the Y and Z directions but a dual Doppler mass also occurs in the X direction. The equations for the Doppler masses are similar to radar frequencies. However, the Doppler mass equations have a constant center frequency that is invariant while the Doppler radar transmitter frequency varies with velocity as per Einstein's time equations. For Doppler radar we get two frequency variations. For a transmitter upon a plane we get:

$$f_L = [C/(C+V)] \, f_O \, [1- (V/C)^2]^{1/2} \tag{20-7}$$

$$f_R = [C/(C-V)] \, f_O \, [1- (V/C)^2]^{1/2} \tag{20-8}$$

In Equations 20-7 and 20-8, the center frequency (f_O) aboard the aircraft is slowed by Einstein's formula $[1-(V/C)^2]^{1/2}$. The frequency to the right is the combination of the Einsteinian decrease in airplane transmitter clock and the Doppler increase $[C/(C-V)]$. For the left frequency, we have both an Einsteinian clock speed drop, and a Doppler drop $[C/(C+V)]$.

The Doppler mass is slightly different. As we speed an object up, the gravitational mass in the X direction becomes more inertial and less gravitational. It is the reverse of what happens to the photon as we slow the photon. Thus:

$$M_{Xg} = M_O \, [1- (V/C)^2 \,]^{1/2} \tag{20-9}$$

In equation 20-9 we see that an object with rest mass of Mo has a decreasing gravitational mass in the X direction as the object speeds up. Thus less and less energy is required to bring an object to higher and higher speeds as we move in free space. This is due to linear space-time theory. It is not what happens in the cyclotron, where orbital steady state space-time theory applies. The gravitational masses in the Y and Z directions for an object moving in the X direction in free space are:

$$M_{Yg} = M_{Zg} = M_{Xg} \, / \, [1-(V/C)^2]^{1/2} \tag{20-10}$$

In equation 20-10 we see that the mass of an object in the Y or Z direction which is moving in free space in the X direction is larger than the mass of the object in the X direction by the Einsteinian formula. We see from equation 19-9 that the mass of the object in the X direction drops with velocity. Thus the total mass of an object in the Y or Z direction is:

$$M_{Yg} = M_{Zg} = M_O \tag{20-11}$$

In equation 20-11, which is for the constant Y/constant Z solution, we see that the mass of an object in free space in the Y or Z directions is invariant. Let us now return to the X direction. In that direction, the mass has dropped. However we have the Doppler masses to consider. Thus:

$$M_L = [C/(C+V_B)] \, M_{Xg} \tag{20-12}$$

$$M_R = [C/(C-V_B)] \, M_{Xg} \tag{20-13}$$

We see in Equation 20-12 that the Doppler or inertial mass to the left is smaller than the gravitational mass. The gravitational mass is decreasing with velocity and the Doppler mass to the left is also decreasing with velocity. In Equation 20-13 we see that the Doppler mass to the right is increasing and this is stronger than the drop in the gravitational mass. The total mass is the inertial mass in the X direction. It is a combination of gravitational mass and photon energy.

The solution for the Doppler mass is a complex spectrum of frequencies and harmonics. If you add up all the neutrons, protons, and electrons, you get a spectrum of frequencies and a Fourier series. The describing function approach produces a single frequency equation that gives the salient points of the solution. The inertial mass in the X direction is the root mean square of the Doppler masses. Thus:

$$M_{Xi} = M_{Xg} \, / \, [1 - (V/C)^2]^{1/2} \tag{20-14}$$

Substituting Equation 20-9 for M_{Xg} we get:

$$M_{Xi} = M_O \tag{20-15}$$

We see that the inertial mass in the X direction is invariant from a single frequency describing function solution. As we move toward light speed we will get some distortions. However, we can now solve the basic problem of the energy required to bring an object up to light speed. Since the mass is invariant in all directions but keeps changing from gravitational mass to inertial mass in the X direction, the solution is similar to classical physics.

$$E = \quad M_O C^2 + \tfrac{1}{2} M_O C^2 \qquad\qquad (20\text{-}16)$$

We see that it takes approximately half the rest mass in energy to bring an object up to near lightspeed. Once we almost achieve light speed, the distortions are a great problem. Additional energy equal to $\tfrac{1}{2}M_{OC}^2$ is necessary to account for the harmonics. For space travel, a trade off is necessary for safety verses time and distance since at present we cannot survive the distortions.

In orbital space-time, the mass keeps doing the same circular track over and over again. This builds up an image of the object. In the case of the cyclotron we keep pumping energy into the machine and forcing the electron or proton to move in circles. We pump up the mass and produce a huge line of charge. Einsteinian space-time is also true for the Bohr orbit. Thus for purely orbital problems:

$$M_{Yg} = M_{Xi} / [1/(V/C)^2]^{1/2} \qquad\qquad (20\text{-}17)$$

In equation 20-17 we see that the gravitational mass in the Y direction is equal to the Einsteinian formula based upon the total or inertial mass in the X direction. Thus:

$$M_{Yg} = M_O / [1\text{-}(V/C)^2]^{1/2} \qquad\qquad (20\text{-}18)$$

We see that for orbital motion the gravitational mass becomes quite large when we approach the speed of light because the perpendicular mass resists the turning. In linear motion, the X masses resist the motion and these eventually turn into pure photon energy, which no longer resists.

The photon has zero gravitational mass in the X direction. It is very soft. In the perpendicular direction, it has mass and is relatively hard. It is hard to bend a light beam. However, a light beam does possess mass and does bend under the gravitational field of a sun. The photon is the purest case. The masses are:

$$M_{Xi} = M_O \qquad\qquad (20\text{-}19)$$

The photon has all inertial mass in the X direction. This makes it soft. The gravitational mass in the X direction is:

$$M_{Xg} = 0 \qquad\qquad (20\text{-}20)$$

We see that the photon has no gravitational mass in the X direction. The gravitational mass of the photon in both the Y and Z directions is:

$$M_{Yg} = M_{Zg} = M_O \qquad\qquad (20\text{-}21)$$

The energy of the photon is:

$$E = M_O \, C^2 \qquad\qquad (20\text{-}22)$$

The momentum of the photon in the X direction is:

$$P = M_O \, C \qquad\qquad (20\text{-}23)$$

We see that when we bring a mass up to light speed in linear space-time, it becomes light. Light speed travel for future man has a lot of obstacles to overcome. If we could bring a spaceship up to the speed of light, it would become light. In general this would cause it to disintegrate unless future man can devise protective fields to prevent self-destruction.

Section 20-2 Linear Space Time

Let us now start the analysis for the properties of linear space-time. The great work of Einstein advanced the thinking of man into the realm of space and time. However there was a lack of understanding of the differences between gravitational mass, inertial mass, and Doppler mass. The problem lies in a restricted singular interpretation of the Michelson/Morley experiment. This interpretation works well for steady state orbital motion but is no good for ordinary classical linear motion.

Now let us look at the Michelson/Morley Experiment again. From the work of Lorentz and Einstein, two postulates were specified:

Postulate 1, the Principle of Relativity stated that the laws of physics are the same or invariant in all inertial systems. The mathematical form of the physical laws remains the same. (Einstein)

Postulate 1 by Einstein is basically true for reference platforms moving at low velocities compared to the speed of light. However in general it is invalid. We live in a nonlinear universe and the laws of nonlinear physics apply. Our Earth moves slowly so the simple describing functions approximate the truth. If we move at higher velocity, the higher harmonics of the nonlinear expressions take over more strongly. At low speeds most harmonics will exist but at higher speeds some harmonics will become as strong as the fundamental. For example, the first Bohr orbit produces a fundamental frequency but high velocity turns this into a Fourier series. Since there are billions of atoms in any mass, we get a spectrum of Fourier series.

The equations of motion and the relationship between large objects moving at very high speeds with respect to each other becomes very difficult to predict. All we can really do is define the motion of high-speed objects with respect to the Earth since it moves slowly in the universe.

Einstein looked at the universe from a mathematical perspective. He took a geometric mean approach. This is pretty good since the Earth is moving very slow compared to the speed of light. His basic answers are true for many areas of space. He felt that every place in the universe was the same but that is not true. There are many places that are quite different than we know. There are many non-linear areas of space-time where human life is not possible.

When two objects are moving toward each other at high speeds, nonlinear electrical theory applies. Objects moving near light speed have different relationships with each other than objects moving slowly compared to the speed of light or objects moving at high speed with respect to an Earth.

There is a big difference between an object in pure free space under the influence of its own tiny gravitational field and an object on planet Earth with the Earth's gravitational field. The laws of pure free space are more like classical physics than Einsteinian relativity. In spite of this, Einstein got excellent results for orbital problems. The main concern of this book is high-speed space travel to our sister planets and nearby stars. We want to achieve at least 50% of light speed. Linear space-time is the most important part of general space-time of which Einstein's orbital space-time is only one particular solution.

Postulate 2 by Lorentz/Einstein stated: The speed of light in vacuum is a constant, independent of the inertial system, the source, and the observer.

Postulate 2 is only partially true. Constant velocity light speed is a very simple solution. It is true that the ideal light speed in pure free space is basically constant over the entire cycle time of the universe. However, this is light speed and not photon speed. The photon speed depends upon the media. It depends upon the motion of the receiver and the sender of the photons. It depends upon the strength of the gravitational fields. It depends upon whether the gravitational fields are moving or relatively stationary. The speed of light is quite variable.

In general, every cubic meter of vacuum in the universe is different than every other cubic meter. The intensity and direction vectors of the electric, magnetic, and gravitational fields in every cubic meter of space in the universe are different. In addition, every cubic meter of space contains an image of the entire universe. The net result is that the light speed in vacuum is different everywhere in the universe.

In Section 20-1, we saw that the mass of the photon in the X direction becomes gravitational mass as it goes from the speed of light to basically zero linear speed within matter. We can calculate the amount of photon energy per unit mass verses velocity. In general, if we took a single photon and reduced its speed in a dielectric, it would become part photon/part mass. The relationship will vary with the velocity squared as compared to the speed of light squared. The relationship was previously stated in Equations 20-14 & 15. Rearranging the terms we get:

$$M_{Xg} = M_O [1-(V/C)^2]^{1/2} \qquad (20\text{-}24)$$

$$M_{Yg} = M_{Zg} = M_O \qquad (20\text{-}25)$$

In Equation 20-24, the gravitational mass of the photon in the X direction varies with its velocity. When it is moving at the speed of light, it has only inertial mass but no gravitational mass. It appears mass-less in free space in the X direction. It still has mass in the Y and Z direction as shown by equation 20-25. Stars will attract it.

The photon is an independent inertial guidance system. Without any forces acting upon it, it will travel at the speed of light. Once you add the synchronized AC fields within a dielectric to the photon, it will slow. Likewise the strong gravitational field of a star will slow the light as U_0 and ε_0 both increase near the star and it will bend more readily around the star. It bends because it is pure gravitational mass in the Y and Z direction and becomes part mass/part photon in the X direction as it drops speed.

Let us now look at what is happening in the Michelson/Morley experiment. A test instrument is built with perpendicular arms. Light from the sun enters the instrument and a comparison is made of the travel time of the light beam in both directions. A null results and equations are produced for the instrument.

The motion of the galaxy is common to everything, so this experiment will be independent of the galaxy motion. The same is true of the motion of the solar system within the galaxy. The Sun/Galaxy motion will not appear readily using this test instrument. More sophisticated instruments can be readily built which will pick up the absolute motion of the Earth including the galaxy motion. However for the moment we have a simple instrument independent of Sun and Galaxy motion.

The instrument was calibrated and one arm was faced toward the sun when the Earth was moving fastest in its orbit. The instrument nulled. It didn't matter if a light on a bench or the sun itself was used. It didn't matter if the Earth was moving at maximum speed toward or away from the Sun. The null was independent of the relative motion of the Earth and Sun. The speed of light was thought to be constant. It was felt that all you get is a Doppler red shift or a Doppler blue shift when moving away from the sun or toward the sun respectfully. Einstein then proceeded with the two postulates to

163

produce special relativity, which is acceptable for an orbital solution but quite wrong for linear space-time.

The big problem is that we think in terms of infinite light speed and Einstein and Lorentz made the world quite limited with our light speed as the only possibility. They turned three-dimensional space-time into a simple electrical circuit on a bench. They produced an electrical universe similar to a bench circuit. They didn't account for a possible spectrum of multiple light speed coexisting universes, or a universe which may presently operate at our light speed but which can go from zero light speed to infinity light speed over an infinity of cycles and configurations.

The universe we live in has infinite light speed capability. Simultaneous events at near infinite light speed can and do occur all over the universe. Einstein and Lorentz incorrectly omitted the entire light speed spectrum. If we move toward infinite light speed in our minds, then we can look at what is happening in the instrument without limiting ourselves the way Einstein and Lorentz did.

As we look at the instrument, we find a photon coming from the sun moving at its normal self propelled inertial speed of C. In pure free space this will be at maximum ideal light speed. Through gravitational fields, it will be slightly less than the perfect ideal light speed. This photon enters the Earth's atmosphere and slows slightly due to the air molecules that reduce the electrical permeability/permittivity constants. There is also a small effect due to the interaction of the gravitational field of the Earth. The photon will start to synchronize with the Earth and drop its speed slightly as well. In addition there will be an even smaller effect due to a Doppler gravitational field effect. Nevertheless, the Earth is moving toward the photon with a basic speed of V_E. The differential velocity as viewed from an infinity light-speed reference plane between the Earth and the photon is:

$$\text{Differential Velocity} = C + V_E \qquad (20\text{-}26)$$

Equation 20-26 is what we would see in classical physics. It is also what we would see from an infinite light-speed reference plane. When the photon enters the test instrument, it is still moving at the same differential velocity that equation 20-26 indicate. However, the internal AC gravitational field spectrum of the instrument will cause the photon to adapt to it. The photon will develop a mass in the direction of travel with respect to the test instrument and the speed will be less than shown above. In spite of this, the photon will travel to the back of the test instrument and re-bound forward with a differential velocity:

$$\text{Differential Velocity} = C - V_E \qquad (20\text{-}27)$$

In equation 20-27, the photon moved from the rear of the instrument and toward the front. Its relative speed now is less. The total round trip time of the photon from the measuring point to the rear of the instrument and back to the measuring point is:

$$T_X = L_X / (C+V_E) + L_X / (C-V_E) \qquad (20\text{-}28)$$

In Equation 20-28 we see that the round trip time of the photon depends upon the sum and difference of the light speed and the Earth's velocity.

Let us use V for the Earth speed in the X direction. As the photon enters the test instrument, we find the time for the round trip to be:

$$T_X = 2(L_X / C) / [1 - (V/C)^2] \qquad (20\text{-}29)$$

In equation (20-29) we find that the round trip time in the X direction has a correction term which involves the square of the Einsteinian correction factor. Thus:

$$T_X = 2(L_X / C) / \{[1-(V/C)^2]^{1/2}\}[1-(V/C)^2)]^{1/2} \qquad (20\text{-}30)$$

In equation 31-30 the Einsteinian correction factor is shown twice for clarity. Let us now look at the Y direction. As worked out by Einstein, the distance the photon travels in the Y direction is longer by simple triangulation since it is moving along the hypotenuse of a triangle. The corresponding time is:

$$T_Y = 2 (L_Y / C) / [1 - (V/C)^2]^{1/2} \qquad (20\text{-}31)$$

Equation 20-30 and 20-31 are basically the Einsteinian equations. More complicated equations can be produced but since we are looking for a simple describing function for space and time, these equations suffice. Of course in order to see the truth of Einstein for orbital motion and the fallacy of Einstein for linear motion, his equations must be studied. For instrument balance:

$$T_X = T_Y \qquad (20\text{-}32)$$

For this simple case in order for Time X to equal Time Y:

$$L_X / (A.A) = L_Y / A) \qquad (20\text{-}33)$$

Where:

$$A = [1-(V/C)^2]^{1/2} \qquad (20\text{-}34)$$

Thus:

$$L_X = A \, L_Y \qquad (20\text{-}35)$$

Therefore:

$$L_X = L_Y \, [1-(V/C)^2]^{1/2} \qquad (20\text{-}36)$$

In equation 20-36 we see that for all other factors being equal, the length (L_X) of the instrument in the direction of travel (X) is equal to Einstein's formula times the length (L_Y) in the perpendicular direction Y.

The Earth is physically shorter in the direction of travel by the Einsteinian correction factor. Since the Earth has a large mass, everything associated with the Earth will shrink in the direction of travel. If we take the instrument and move it 90 degrees, the X-axis will get larger and the Y-axis will get smaller. It will always maintain the size difference.

Einstein's relativity concerned a mathematical shortening of the X-axis. In true space-time there is an actual shortening of the X-axis and some possible enlargement of the Y-axis and Z-axis as well. True space-time is quite physical and subject to ordinary classical physics whereas Einsteinian space-time is purely mathematical and lacks some physical reality.

There is a time delay in these changes in the X and Y direction during rotation. If we rotate the instrument at extremely high speed and make it from a huge amount of mass, then the readings will change to show the effects of Earth speed verses the null point. Yet the original instrument is not designed to rotate at very high speed nor does it have a very large mass. Other correction factors will occur as well. However, things are quite different in pure free space. Upon the Earth we are locked into the mother gravitational field. Our instrument is only a tiny test instrument. It obeys the Earth's field. The Earth shrinks in the direction of motion and elongates somewhat in the perpendicular direction. We need a very massive instrument to beat Mother Earth's field. On the other hand once we move away from the Earth we will have good results.

In general the distances will change with velocity. The relative speeds of C+V and C-V will apply. Einstein chose one solution. We see that the results are the same for length for the Doppler space-time solution as for the Einsteinian solution, except that for the Doppler solution, the distances are a variable ratio of the arms rather than just shrinkage of the arm in the X direction.

In equation 20-36 we see that the moving object shrinks in the direction of motion as compared to the perpendicular direction. Yet, the perpendicular direction tends to elongate due to the motion. These are physical effects on large bodied objects. The atomic spacing between atoms within the instrument is subject to physical stresses due to motion.

The effects are non-linear but a steel pipe moving in the X direction will be compressed greatly due to the motion. The space-time equations show that not only will acceleration compress an object but also constant motion will produce permanent compression. If you look at the object as a ball of waves, if we move the ball in the X direction, the front waves will get distorted and compressed. The ball will shrink in the X direction and elongate somewhat in the Y direction. Since Einstein used constant Y and Z, this book will do the same so a comparison can be made

Einstein's relativity equations gave the wrong impressions. The moving object is the one that is compressed. The stationary object is not compressed. Our motion changes us with respect to the sun whereas the sun does not change.

Let us now look at the time clock and see what happens with a moving time clock. This will resolve the clock paradox as well. The clock paradox clearly shows the work of Einstein to be incorrect for linear motion. We can see this by looking at the mechanism by which the clock of a moving object slows down

Let us use the same M/M instrument as a clock. Let us keep the photon source within the instrument so that photons are constantly generated and move back and forth in the X-axis and Y-axis while being counted and measured. We then have a fancy instrument clock. The atomic clocks on a smaller scale will work similarly.

To keep the solution simple, let us consider the case where the speed of the instrument is rather low compared to the speed of light. Let us also assume that most of the distortion is in the X axis direction and that the Y axis distortion (bending) in the X direction is basically smaller than the distance traveled in the X direction. Otherwise we must use nonlinear analysis which would apply as soon as the velocity gets too large.

For the simple case, the instrument nulls and the time of travel in the X-axis equals the time of travel in the Y-axis.

$$T_X = T_Y \tag{20-37}$$

We can use the Y-axis as the reference axis. Thus:

$$T_Y = 2(L_Y / C)/ [1-(V/C)^2]^{1/2} \tag{20-38}$$

Let $T_O = 2L_Y/C$, thus:

$$T_Y = T_O / [1- (V/C)^2]^{1/2} \tag{20-39}$$

We see that a moving clock mechanism takes more time to travel internally since the path length is longer due to motion. The clock slows as per Einstein.

Einstein was incorrect when he assumed that a clock on the ground would appear slower to a moving observer in the sky. If we look at a clock on the surface of the Earth and one in a satellite, we will notice that both are distorted and slowed due to galaxy rotation. Both are distorted and slowed due to the solar system motion. Both are distorted and slowed due to the rotation of the Earth.

The Earth is a large object. The rotating clock on the satellite does not disturb the gravitational field of the

Earth. The distortions of the Earth clock and the satellite clock are common mode due to Earth rotation.

Now the clock in the sky has an additional velocity term due to its motion. In general the velocity vectors of Earth motion, sun motion, and galaxy motion are not simple vector additions due to the nonlinear distortions of all the dimensions. In general one must apply a root mean square solution to the velocities. The fact that the satellite is moving in the direction of galaxy rotation or against it does not readily show up since galaxy rotation is common mode to everything. However, more sophisticated rotating instruments with high angular velocity and heavy mass should show some differences.

The clock in the sky is slower than the clock on the ground since it has an additional motion above the common mode motions. They are both slowed common-mode due to all the other motions. If you took radar measurements of this including the Doppler corrections you would find this to be true.

If you were in the spaceship orbiting the Earth, the Earth's clock after Doppler corrections would appear faster. The clock paradox is not a paradox at all. It was just an error in thinking on the part of Einstein. Einstein applied steady state electrical theory to linear transient problems. It was a simple solution but it was only partially true.

Let us take two spaceships with radar system clocks and a ground station with the same clock. If the spaceships took off in opposite directions, they both would see each other's clock as moving the same after correcting for the Doppler effect. This will produce fewer and fewer pulses per second as the ships speed up. They both would see the Earth clock as moving faster after correcting for the Doppler. They could then travel quite a long time and later turn around fairly rapidly. Without considering accelerating and deceleration for a short turn around time compared to the total voyage time, they would start to pick up more clock pulses per second when they moved toward each other due to the Doppler. The time difference in the radar pulse counter would slowly head toward zero. They would still see each other's clock as the same and the Earth clock as moving faster.

Finally when they arrive once more upon the Earth, the two spaceships would find that the measured clock ticks of their own ship and the radar clock counter measuring the other ship are the same. The clocks would read exactly the same when they arrive upon the Earth. However the Earth clock would be faster.

Two ships traveling in opposite directions each see the other's clock as identical for identical speed. Of course in this analysis we assumed that the effect of the gravitational field of the Earth did not change the spaceship's clock. If we move far enough away from the Earth, we will get a change in the root mean square factors in the initial distortion of the clocks. There will be a tiny increase in clock speed due to being apart from the Earth. However the big distortion factor is the galaxy speed and this cannot be readily escaped. Yet, this is common mode to everything and all instruments tend to be distorted by it and it nulls out.

Let us now look at the mass of an object as it varies with velocity. In the original theory by Einstein, the relativistic mass of an object was computed using two balls that collided.

The first ball was on a stable reference frame such as the Earth and moved upward with a velocity V_{YU}. Forgetting about gravitational effects, a second ball was on a moving platform above the Earth. The platform was moving to the right with a velocity V. An identical ball was sent down from the moving platform at a downward velocity V_{YD}.

A perfect collision between the two balls occurred and the momentum in the Y direction was identical. The upward ball moved a distance upward of Y_O and the downward ball moved an equal distance. Since the momentum was conserved:

$$M_U V_{YU} = M_D V_{YD} \qquad (20\text{-}40)$$

In equation 20-40 Einstein reasoned that the upward momentum in the Y direction was equal to the downward momentum in the Y direction since both balls traveled an identical distance. In addition, both balls stopped at the halfway point and returned to their reference frames in an identical manner from when they started. The lower ball merely went up and down while the upper ball moved to the right while moving up and down. When we look at the lower ball we see that it moves a distance of only $2Y_O$ in a time T_O. The momentum of the lower ball is:

$$MV = M_U Y_O / T_O \qquad (20\text{-}41)$$

Things are a little different for the upper ball. It travels a much larger distance since the reference platform was moving to the right with a velocity V. Likewise it can be said that it travels the same distance in the Y axis direction but it takes longer to get there. The momentum then becomes:

$$MV = M_D Y_O / T = M_D Y_O / \{T / [1 - (V/C)^2]^{1/2}\} \qquad (20\text{-}42)$$

By equating the momentum of Equations 20-41 and 20-42, Einstein concluded that:

$$M_D = M_U / [1 - (V/C)^2]^{1/2} \qquad (20\text{-}43)$$

Equation 20-43 is Einstein's famous equation for the variation of mass with velocity. The rest mass is the upward mass and the moving mass is the downward mass.

The problem with Einstein's calculations was that it made no difference which was the reference frame and which was the moving frame. Each object according to Einstein would look at the opposite moving object as the relativistic mass. This is not correct. His universe was a strange mathematical non-physically realizable universe.

If you add energy to an object, its inertial mass increases. If you make it go faster it will have a larger mass. If you just look upon another object which you did not add energy to, it won't have a higher mass. Einstein's equations for special relativity are false for linear motion. Einstein's solution is correct for orbital motion where the reference is stationary such as the cyclotron or for planetary motion when the sun is used as the reference. As long as the reference is basically stationary, Einstein's equations are the best mathematical fit.

Einstein's equations are true when we consider the Earth as the reference platform. The Earth motions distort everything upon the Earth. Everything tends to be common mode. The moving clock above the Earth will slow. However our clock will look faster to it. The reason is that the Earth and the clock have been adjusted for galaxy motion, sun orbit, etc. The only thing different is that the clock above the Earth is moving with respect to us. That clock has already been adjusted to match the Earth. Thus Einstein's equations are correct as long as this Earth is the reference platform. The little clock only has a very tiny gravitational field. Therefore it is dependent upon the Earth's gravitational field for corrections.

It is also important to note that Einstein got the correct mass equation for the Y-axis with respect to the total mass on the X-axis. His orbital mass equation is true. However, he failed to recognize that moving a mass at the speed of light converts gravitational mass in the X direction into inertial mass.

Section 20-3: Doppler Mass Center of Gravity

Let us now look at the concept of the Doppler mass center of gravity. Let us investigate a huge ball that is perfectly spherical moving in an area of free space between galaxies. If the ball is perfectly stationary in the universe, it will be perfectly round.

A three-dimensional Doppler laser gyroscope within the ball will always null and it will have an offset error if moved in any direction at any velocity. When initially perfectly stationary, the device can be calibrated by moving it in each of three planar rotations to see that it is equally balanced. Later, when moved, it can be calibrated for linear motion. The accuracy of the instrument depends upon the magnitude of its own gravitational field as compared to the gravitational field of the surroundings. An ideal instrument would be heavy and would be between galaxies. It should work well with a 10/1 ratio of self-gravitational field to that of the surroundings. An ideal ratio would be 10,000 to 1 for near perfect results.

The gravitational field of the perfect ball will extend outward for millions of miles producing a huge electromagnetic ball. It will be perfectly spherical. The ball and the gravitational field will both be perfect. However, the field is a spectrum of individual atomic fields. It is a perfect but highly complex sum total of billions of individual fields operating at a spectrum of frequencies.

172

Let us now push the ball and let go. Initially the ball will compress at the point we pushed it. This will move all over the ball and to the gravitational field itself. Later when we achieve stability, we will find that the ball is permanently distorted in the direction that it is pushed. It will also show some distortion in the perpendicular direction as the pressures equalize. It gets shorter in the direction of motion but longer perpendicular to the direction of motion. The gravitational field will change as well. It will react outward from the push at the speed of light.

The inertia of the object is caused by the displacement of all the current loops in the present from the past and the force generated by Ampere's law. In addition, the atomic spacing will be affected causing shrinkage in the direction of travel.

The pushing force will compress the ball in the rear as a transient condition. Later in steady state the front will compress and the rear will spring back toward normal. A permanent shrinkage will occur.

Eventually, the internal gravitational field will have compressed waves in the front of the motion and elongated waves in the rear of the motion. The ball will then have a Doppler mass distribution. The density of the mass will be greater in the front and less in the rear. The Doppler mass equations are:

$$\text{Mass front} = M_X \, C/(C-V) \qquad\qquad (20\text{-}44)$$

$$\text{Mass rear} = M_X \, C/ \, (C+V) \qquad\qquad (20\text{-}45)$$

The Doppler geometric-mean mass, which is the inertial mass, calculates to be:

$$M_{Xi} = M_X \, / \, [1- (V/C)^2]^{1/2} \qquad\qquad (20\text{-}46)$$

Let us look at the case where the ball is moving rather slowly compared to the speed of light. We can then approximate the front and rear Doppler masses as follows:

$$M_F = M_X \, [1 + (V/C)] \text{ approximately} \qquad (20\text{-}47)$$

$$M_R = M_X \, [1-(V/C)] \text{ approximately} \qquad (20\text{-}48)$$

If we look at the front mass at a distance of (+L/2) from the center of the ball and the rear mass at a distance of (-L/2) from the center, the new center of mass is:

$$\text{Center of Mass} = +L \,(V/2C) \qquad (20\text{-}49)$$

In equation 20-49 we find that a body moving at modest speed has a shifted center of mass equal to half the length of the body times the ratio of the velocity of the moving body to the speed of light. This is for a body such as our Earth that moves slowly compared to the speed of light.

The faster the body moves, the closer the center of gravity will move toward the front surface. For high-speed objects, we would get another Einsteinian type equation with Doppler corrections for the center of gravity of the object. Yet for our Earth and most objects, equation 20-49 will suffice.

If we look at our Earth we will find that the axis of the Earth will be off centered from the true center of gravity. This in turn is offset from the physical center due to an uneven distribution of mass. The axis of rotation of the Earth is off centered from the stationary center of gravity due to the motion of the Earth. There is a Doppler shift as per Equation 20-49 in the center of gravity of the Earth.

The velocity of the Earth increases and decreases as it orbits the sun. The center of gravity will shift every day. This will cause the Earth to tilt. We get summer and winter due to the Doppler shift of the center of gravity of the Earth.

The change of the center of gravity of Mercury should be more severe as it moves closer to the sun. Einstein calculated the motion of Mercury from his equations whereas no one else could. His space-time equations have always been a good fit. However, Einstein never brought to mind the Doppler mass effect. Although his equations are good, the real reason for the motion of this Earth and Mercury as well is that the center of gravity changes as the velocity of the planet changes. In addition, changes in position toward or away from the sun result in perpendicular forces as the gravitational fields interact with each other.

Our moon is phase locked to us like a synchronous motor due to the offset center of gravity. The galaxy motion causes center of gravity distortions in our sun, our Earth, and our moon. However, this may not be easy to detect since the radius around which the distortions are occurring is so far from us. Since both the Earth and sun are moving within the galaxy, it may not be so evident that the Earth's axis has an offset component due to galaxy motion.

Our moon will have an offset axis due to our motion around the sun and its motion around us. We appear phase locked to the face of the moon. If the moon was perfectly uniform and there was no Doppler shift, the moon should rotate. However the moon is not quite uniform. It is subject to a Doppler shifts, due to rotation around us and around the sun. The forces acting upon the moon tend to produce a phase locked condition after awhile.

The Doppler shift in the center of gravity of the Earth also tends to add heat energy to the core. The energy of rotation is slowly turned into additional heat in the core of the Earth. The Earth very slowly moves toward the sun over a long period of time. The rotation of the Earth also tends to slow. Eventually in the far future one face of the Earth will end up facing the sun. Then the Earth no longer supports life, as we know it.

Section 20-4: Doppler Relativity reference System

Let us look at an object such as a perfect sphere that is rotated upon an axis until it reaches near the speed of light at its circumference.

Einsteinian relativity found that a man in an elevator moving at a constant velocity V could not detect his motion. Yet, if the man were spinning with a constant rotational velocity, he would be pinned to the outer surface of a round or spherical elevator. Rotational motion can be detected but not linear motion as per Einsteinian relativity. This is a fallacy.

True Doppler relativity does not distinguish between linear motion and rotational motion. Both can be detected. Rotational motion is easier to detect and linear motion is harder to detect. A three-dimensional Doppler gyroscope set in pure free space will measure velocities elsewhere. It will not null upon this Earth, however once in our gravitational field, it will become distorted.

Perfect instruments must be kept away from strong gravitational fields; otherwise they will exhibit errors. It is difficult to overcome the distortions of the Earth itself. Once we null the instrument here on Earth, it will tend to always null. However, if we null it in outer space then it will not null here.

In a universe where all the planets and stars are moving and the entire universe may be rotating as well, a reference point is hard to find. Of course for a possible multi-light-speed set of coexisting universes, we can use the highest light speed universes as the reference. We can imagine our universe defined in terms of distances and time upon the highest light speed references. However for ourselves we need to know what the reference is for our own universe.

The basic reference for the universe is speed. If we take an object in pure space far from any galaxy or between galaxies, the forces upon the object will be nearly zero. Let us take a sphere and place it within a larger sphere that reflects the light back to the inner sphere. Since we are far from any galaxies, we exist in perfect space. Let us beam light in all directions from the inner sphere to the outer sphere and obtain a null by firing little rockets on the outside of the sphere until we get a perfect null pattern. When the Doppler signals in all three axes, return with a perfect null, the spherical spaceship is perfectly stationary.

As we move the spaceship with a velocity V with respect to a ruler on the spaceship, we will find that the null is destroyed. If we try this upon the Earth, we get problems from gravitational distortions due to the motion of the Earth, since we are the Earth and not the instrument. The perfect instrument will have corrected lengths that tend to negate easy measurements. In perfect space we do not have problems with nearby gravitational fields. For an instrument to work properly, the gravitational field of the instrument itself must be much larger than the external gravitational fields of the surroundings. A ratio of ten thousand to one would be ideal. However instruments should work with ratios as little as ten to one.

The M/M experiment exists within the Earth's gravitational field and lacks the ability to test for zero speed. The mass of the test instruments must be free of external influences. This is clearly an impossible problem for experiment upon this Earth. The best we can achieve is experiments on a space ship at a gravitational null point between our sun and another local star. Everything else is too far to readily accomplish.

Our spaceship can be nulled to obtain perfect rest. Then we can measure the Doppler null shifts to determine our speed. As long as we are moving slowly compared to the speed of light, we remain fairly linear.

As we speed up, our spherical spaceship will flatten in the direction of motion and expand perpendicular to the direction of motion. We will become distorted. The solution to the Michelson/Morley experiment shows that the ratio of the two perpendicular axis changes as the object moves in orbit.

The distortions and the Doppler null shifts tell us our speed. Thus Doppler relativity shows distortions as speed increases. It also shows that speed itself is the reference in our universe.

If you look at Einstein's equations you will find that they all include the velocity V in them. Distance and time change but V remains the same. Thus, the true reference system is the velocity V.

The primary principles of Doppler relativity state that:

Doppler Relativity Principle 1:

All non-rotating linear reference systems moving in pure free space with the same absolute velocity (v) are equivalent.

Principle number one states that all linear non-rotating reference systems moving with the same absolute velocity are equivalent. The rulers, the distortions, and the time clocks of these reference systems will be identical. If the two systems are moving in the same vector direction, radar signals between the two will show perfect clock frequencies. If the two systems are moving in opposite directions, the Doppler clock measurements will be lower identically for both systems and spaceships. If the two spaceships are moving toward each other, then the clock frequencies measured will both be higher.

If you have one space ship with a particular absolute velocity and another with a different absolute velocity, the two systems are not equivalent. Both will have different rulers and clocks. If the velocities are reasonably slow relative to the speed of light, then we can calculate the various rulers and clock frequencies and obtain the Doppler radar calculations. These will require more complex calculations than for equivalent systems.

If one space ship is moving at half the speed of light, things get very complicated, as the distortions are very large. The error that Einstein made in relativity is that he assumed that the speed of light was the only reference. The speed of light is one reference and the absolute speed of an object is the second reference.

Let us now look at a rotational object. We see that all reference systems that move at the same absolute velocity are equivalent. This is difficult to produce except between galaxies. However, if we look at two galaxies which are moving at the same rotational speed and look at two planet Earth's which are in the same location, then we can say that the two planet Earth's will produce identical results. In general except in free space it is difficult to find two objects with identical speeds.

Of course when all objects are moving slowly compared to the speed of light, we get good results by assuming that one object is stationary. Let us now take a solid ball far out in space and start to rotate it. Far in space, all objects rotating with the same circular momentum are identical if their linear velocities are the same. Thus:

Doppler Relativity Principle 2:

All reference systems moving in pure free space with the same linear velocity and the same rotational velocity are identical.

Rule No 1 was for non-rotational linear systems whereas rule No.2 includes rotations. Of course if we are on a fast moving rotating spaceship we will know it pretty rapidly. The human body can take steady linear motion up to nearly fifty percent of light speed without problem. However, once we rotate too fast, we will just perish.

Let us rotate the perfect sphere. As we rotate the sphere, it will get longer perpendicular to the axis of rotation. It will also start to shrink along the axis. This is just simple strength of materials mechanical theory. We get distortions similar to the Michelson/ Morley experiment within the ball.

As we add more and more rotational photon energy to the ball, it will get longer perpendicular to the axis of rotation and it will flatten at the poles. The mass of the ball will continue to rise. A clock within the ball will slow as per an equivalent Einsteinian or root mean square Doppler time formula. The length along the axis will shrink as per a similar Einsteinian formula.

Although the length along the axis shrinks, the diameter expands greatly. During the rotation of a perfect sphere we find that both the axial and transverse dimensions change.

As the speed of the rotating sphere approaches nearly the speed of light on its circumference, its diameter expands greatly and it's inertial mass approaches a large number. In addition, its length approaches zero.

We see how the rings of Saturn were formed and how galaxies can move from flat planes to more spherical shapes. If you take energy moving close to the speed of light in a circular plane, it will look like a flat disk spinning rapidly. As the disk slows, it will start to form a spherical shape in the middle. As we move further away from the center, the velocity will approach the speed of light. As we move close to the center the velocity will approach zero.

If we spin the Earth faster and faster, it would flatten at the poles and elongate perpendicular to the axis of rotation. We would then produce rings around our Earth. As the Earth slowed, the rings would either turn into a moon or several moons or remain as rings for awhile.

Very high rotational forces formed Saturn. The same was probably true of our solar system. Since it operates in one plane, it probably was formed from a large ring around the sun moving close to the speed of light. Yet, many alternate possibilities exist. The only thing important is that a large spinning disk can turn into a sun and planets. Photon energy spinning in a plane can become a solar system or a galaxy for that matter.

Einstein's relativity turns everything into a simple problem. All reference frames are equal. Yet, that is not true. The only times things are equal is when we have identical rotational velocities and identical axial and transverse velocities. Of course we must then be free of any major stationary gravitational fields or moving gravitational fields.

The only conclusion is that all reference planes in the entire universe are unequal. Here and there you will find reference planes which are almost equal to each other. Even upon the Earth, you will not find two reference planes that are absolutely equal. The closest you can get is two reference frames in the same laboratory Thus we get:

Doppler Relativity Principle 3:

No two reference planes in the entire universe are absolutely identical.

Unfortunately we cannot find two reference planes in the entire universe which have the identical conditions. Even if you could find two spaceships with the exact rotation and linear velocity, they will still exist with gravitational fields near them that are moving and different.

Einstein though that all reference-systems were identical and could be tied to the speed of light. In truth, we live in a universe where you can find nothing absolutely identical. The best we can get is areas of pure free space far from stars and the centers of galaxies. On the other hand, this planet Earth moves relatively slowly so, it is a great reference itself.

Section 20-5 Length & Time

In Doppler space-time, mass is invariant. Although there are changes from gravitational mass in the X direction to inertial mass, the absolute mass remains the same. Let us now look at length and time in Doppler space-time.

For the simple linear solution, L_Y and L_Z remain constant. For this solution, the only thing that changes is the length in the X direction or L_X. Let us look at the total Doppler length in the X direction. To the right and left it is:

$$L_{XR} = [L_0 [1-(V/C)^2]^{1/2}] . [C/ (C+V)] \qquad (20\text{-}50)$$

$$L_{XL} = [L_0 [1-(V/C)^2]^{1/2}] . [C /(C-V)] \qquad (20\text{-}51)$$

The RMS value of the Doppler is:

$$L_{XRMS} = L_0 \qquad (20\text{-}52)$$

As the velocity reaches light speed, we see that the Doppler length to the right moves toward zero and the Doppler length to the left moves toward infinity, while the RMS Doppler is steady at L_0. The image of an object stretches out toward zero in the forward direction and stretches out toward infinity in the rearward direction but the object remains the same size from a root mean square describing function analysis.

We see that length is invariant in true space-time. Both mass and length are invariant and quite classical for linear space-time. Now let us look at the time clock. The time to the right and left are:

$$T_{XR} = [T_0 [1-(V/C)^2]^{1/2}] . [C / (C+V)] \qquad (20\text{-}53)$$

$$T_{XL} = [T_0 [1-(V/C)^2]^{1/2}] . [C/(C-V)] \qquad (20\text{-}54)$$

The RMS Doppler time is:

$$T_{XRMS} = T_0 \qquad (20\text{-}55)$$

We see in equation 20-55 that the RMS Doppler time is also invariant. Therefore space, time, and mass are all invariant in true linear space-time for the constant Y/constant Z solution. Einstein's solution for Orbital space-time is not applicable to linear space-time. Only classical techniques are proper.

Section 20-6: Charge Invariance

Let us look at the invariance of charge with velocity. The electrical charge Q of the electron will change when it moves. It will experience a Doppler effect. This will give rise to the magnetic field of the electron. Just like mass and length the Doppler equations for charge are:

$$Q_{XR} = [Q_0 \, [1-(V/C)^2]^{1/2}] \cdot [C \, /(C-V)] \qquad (20\text{-}56)$$

$$Q_{XL} = [Q_0 \, [1-(V/C)^2]^{1/2}] \cdot [C/(C+V)] \qquad (20\text{-}57)$$

$$Q_{XRMS} = Q_0 \qquad (20\text{-}58)$$

$$Q = Q_0 \, / \, [1-(V/C)^2]^{1/2} \qquad (20\text{-}59)$$

We see in equation 20-59, the build up of charge for orbital motion. The faster the electron moves, the greater the charge builds up. Of course we define this effect by the magnetic equations. The magnetic field is the result of the Doppler Effect on moving electrons. In addition, it is the charge buildup due to repetitive cycles of the electrons.

For linear motion, most of the magnetic effect is spread at the front of the electron and behind the electron. The perpendicular charge of the electron in pure free outer space tends to remain fairly constant just as the mass does. There is always some variation however.

Once we move a charged particle, we get a Doppler magnetic field. The magnetic field is merely the Doppler effect of the electric field.

Chapter 21: Space Travel

Section 21-0 Introduction

One of the hopes and dreams of man is to be able to travel the stars in search of new Earths to inhabit. As we look around our area of the universe we do not find any viable Earths in our vicinity. We are trapped on planet Earth and the only escape is by means of the transmission of our souls to new Earths elsewhere in the present universe or upon higher new Earths in the future universe.

As we move upward in light speed, higher man will have much longer lives. This will give higher man the ability to travel the stars while we cannot do that readily. It may very well be that for higher man God has provided multiple Earths upon a single star. He will be born upon one Earth and have other Earths to reach.

It is difficult for us to travel to Mars or Venus. Yet we can do it someday. The biggest problem is the necessity for an engine that can use atomic energy for propulsion. Although the distances in space are quite large both for us and higher man, it is necessary to travel to local planets.

The energy of the protons and subparticles was locked into them at the first big bang. Some of this energy was locked into the protons during the explosion of forming galaxies. Although it was originally thought by me that somehow we could unlock the energy in the proton to produce photons, during the last few weeks I have come to understand the principle of three universes, and I realize that it is basically impossible to do.

Photons tend to be free energy. They do not exist simultaneously in the past and the present; not our photons anyway. God's photons coexist with all dimensions. This enables God to do the most amazing things if God shall so choose.

It appears that there are different levels of the future universe and the past universe. The very split second future and split second past works within the Bohr atom. Then if we move a little further in time we achieve an entirely different but quite similar universe. Therefore there is immediate interaction between the past, the present, and the future. There is also interaction in which the universe produces similar but separate results. For example a black hole exploding in a future universe has only a little effect upon our universe.

These complications cause the protons to be locked into space and time for small steps in time while the future protons decrease in energy levels for large steps in time. It appears that we cannot obtain energy from the proton to fuel our spaceships of the future.

The only way we can achieve near light speed is to produce atomic fuels that will give us photonic energy from a limited amount of mass. There is not enough fuel that could raise a spaceship up to the speed of light. Although for orbital motion Einstein's equation hold true, ordinary linear motion in free space acts more like classical physics. Once we rotate an object in orbital motion, there is energy built up which brings it into a very high energy level.

In spit of this we could reach perhaps 0.5C. Let us assume that to be the case. Therefore let us study what happens when we build an atomic engine and take a spaceship into outer space up to 0.5C. First we need to produce the atomic fuel.

Section 21-1 Atomic Fuel

One solution is to produce a controlled miniature atomic or hydrogen type bomb. It would be necessary to control the reaction by use of electromagnetic fields and laser type rays. We could then focus the beam to the rear and the spaceship will rise on the photonic beam. Unfortunately we only use a little of the atomic mass and we would not achieve too high a velocity.

A more advanced method would be to take photonic energy and modify atoms such that they become part photonic part mass atoms. This would require some fancy chemical compounds which could hold the energy in place. We could then have other compounds which act to destroy the first compound. In this way we pump photonic energy into matter which holds it in place.

The atomic engine would then mix the compounds with the right type of electromagnetic and laser fields to produce photonic fuels. Another method would be to produce super radioactive atomic elements which unleash tremendous amount of radioactive energy. No doubt scientific man upon this Earth will produce one type of fuel or another. For the moment let us assume that we can have enough fuel to reach 0.5C.

Section 21-2: The Atomic Engine

The atomic engine is the most important physical device for the trip. We must be able to travel up to fifty percent of light speed (0.5C). In addition we must maintain continuous acceleration at 32 feet per second per second (1G). Weightlessness will not be acceptable except for short periods of time. Right now how long a person can stay in space in a weightless condition is being studied. This problem will be eliminated if we always keep a spaceship under continuous acceleration.

In order to accomplish 0.5C, we must accelerate uniformly at our rate of gravity of 32 feet per second per second. This will take approximately 6 months. Then we must slowly turn around and slow for 6 months in the reverse direction to return to the Earth.

The ship will have two main atomic engines. The rear one will bring us up to one-half light speed. The front one will protect us against anything in our path.

As we travel near half light speed, we will produce magnetic and photonic waves in both the front and the rear of us. These will act as protective barriers as well. The front of the spaceship must come to a point, which provides a degree of protection when used in combination with the protective waves we produce. If we find a rock ahead of us, we must increase the energy of the front atomic engine and the rear atomic engine in synchronization. The total power will increase but the crew will still experience a force of only one G. The front atomic engine will be focused into a ray that completely destroys the rock and everything will be safe.

Of course we cannot destroy a huge space rock. Therefore we must always look ahead and see what is there. If a space rock is reflecting sunlight, then we can visibly see it. If a dark huge rock is in our path, we will not see it. Our front atomic waves will rebound from it and our computers will have to decide what evasive action is necessary.

The engine will permit high-speed space travel up to 0.5C without the necessity of weightlessness. Ordinary people can ride the spacecraft. Ordinary people can travel to Mars. Takeoffs and landings are easy. We can ride the beam up, and ride the beam down when we land.

The atomic engine permits easy space travel. A frail elderly man could travel to Mars. People can be born on the spacecraft. You can leave the Earth at 1.5 G's or less. A man who weighs 160 pounds will weight 240 pounds during takeoff.

Section 21-3: Trip to the Moon

The trip to the moon tests out the spacecraft design. It gives us the opportunity to take off from the Earth and to land right back upon the Earth. We will come down backward riding the beam downward while gravity pulls us down. The front beam will also be used to prevent the spacecraft from tilting.

The spacecraft can land on any hard rocky surface. It could also melt various surfaces and produce it own landing floor. For the Earth one or more sites will be used and precision corrections must be made during the spacecraft flight. We could bring the spacecraft to a perfect stop thousands of miles from the Earth and watch the Earth rotate until we are ready for a landing.

The time to reach the moon and the peak velocity is found by the simple physics formulas for constant acceleration. They are:

$$S = \frac{1}{2} At^2 \tag{21-1}$$

$$V = At \tag{21-2}$$

The moon is approximately 240,000 miles away. Half the distance is 120,000. The time to achieve half the distance is found by equation 21-1. This amounts to 6290 seconds when the acceleration is 32ft/sec^2. It takes 1.75 hours to reach half the distance to the moon. The speed is found from equation 21-2 to be 137,400 miles per hour.

At the halfway point, the front engine must produce a deceleration of 1G. The rear engine will be reduced to a small value to maintain stability. It will take another 1.75 hours to reach the moon. The entire trip will take 3.5 hours in comfort. Of course it will take time to turn around and land. At least an additional hour is required during landing.

The only discomfort will occur during the changeover from acceleration to deceleration. The people will become weightless for a few seconds and the seats will be turned around automatically. The people must be strapped into their seats during that time. The entire inside of the ship will become the mirror image as far as the passengers are concerned.

Prior to landing, the ship must turn around. Another period of floating will occur. All these periods should take only a few minutes. The crew of course will be quite used to weightlessness for short periods of time. However some will have to be trained to work under weightless conditions for emergency repairs and the like. The passengers will merely have to remain strapped in their seats during this time. A special room could be set up for those passengers who wish to experience weightlessness for awhile.

The trips to the moon will become common place. It will take no longer to get to the moon then it does to go from New York to California. Therefore Hotels will be set up on the moon. Everything must be brought in but the atomic engines will enable solid rock to be cut and fused together. We will then be able to build domes and bring water and supplies with us. There will be constant daily trips to the moon. People will honeymoon on the moon as well.

We will develop space stations on the moon with sufficient supplies to keep several hundred people alive for up to one hundred years. Cargo ships will bring a lot of water to the moon and we will produce a small city in the rocks of the moon with sunlight coming in. We can then grow fresh tomatoes and other vegetables.

Section 21-4: Trips to Mars and Venus

Mars is approximately 142 million miles from the sun and Venus is approximately 67 million miles from the sun. We are 93 million miles from the sun. Although orbits vary, we can assume that at some time we will be approximately 55 million miles from either of them. Astronomers and the like know the actual distances. However for the discussion here, it is only necessary to understand the approximate distances. The time to reach 27.5 million miles at 1G would be 26.5 hours with a peak speed of 2.08 million miles per hour. The entire trip to Mars or Venus would take 53 hours or 2 days and 5 hours.

Of course, the exact time of the trip depends upon the orbit and the time of year. The trip to the moon could be a day trip. Many people can go. Millions of people can eventually make the trip to the moon each year. It will not be an expensive trip. The trip to Mars and Venus will be much more expensive for the average person.

Section 21-5: Trip to reach Maximum Speed

Our brave astronauts will have to test the ability of man to achieve high speeds and the ability of man to reach deep space. We must test the limits. We must find out how far we can travel from the Solar System and still survive. We will head toward outer space and turn back when necessary.

At the conservative 1G it will take us six months to achieve 50% of light speed and another six months to return to the Earth. The biggest problem besides fuel is the Doppler length distortions.

Let us look at the following chart for the variation of gravitational shrinkage and inertial expansion with respect to increasing light speed.

Velocity	Grav. Length	Rear length	Forward length
0.0C	1.000	1.000	1.000
0.1C	0.995	1.105	0.904
0.2C	0.980	1.225	0.816
0.5C	0.866	1.732	0.577
0.9C	0.4436	4.359	0.229
0.99C	0.141	14.107	0.0709
0.999C	0.022	44.710	0.0224

The Einsteinian gravitational length is the shrinkage of distance with velocity as per the space-time equations produced by Einstein. Unfortunately his relativity did not account for the Doppler differences. The length of an object increases in the rearward direction of motion and decreases in the forward direction of motion. These differences produce distortions which will tend to destroy living creatures.

These differences are not important at low velocities such as we have upon the face of this Earth. When we attempt to bring a spaceship upward toward the speed of light, objects get distorted. At fifty percent of the speed of light a 6-foot person becomes over 10.4 feet in the rearward direction but only 3.5 feet in the forward position. A ceiling of a spaceship compartment, which is twenty feet, becomes only 10.4 feet in the rearward dimension. We must make sure the ceilings are at least 20 feet high between compartments on the spaceship.

When we look at the workings of our own body we find that our organs will struggle really hard when we attempt to reach 50% of light speed. Einstein did not understand modern Doppler radar techniques. He used a root mean square approach to his space-time. This works well for orbital motion but fails to work for ordinary classical linear physics motion.

The human body will start to fail at 0.5C or less. The Einsteinian elevator problem was incorrect. Instruments will sense constant velocities as soon as they become large as we approach the speed of light. The effort to turn a one pound ruler around will be quite noticeable at these speeds especially when the engines are turned off and the spaceship is moving at constant velocity.

The problem with Einstein's solution is that it is an orbital space time solution an not a linear space time solution. This will be clear once we have research laboratories moving at high speed. This does not deny his fantastic work with their excellent orbital space time results.

As we look at the chart we find that a spacecraft traveling at 0.1C has a slight Einsteinian shrinkage of one half of one percent. This is small. The spacecraft looks ten percent smaller in front of it and ten percent ten percent larger to the rear of itself. A man would experience the same distortions. When we go up to 20 percent of light speed, the Einsteinian gravitational shrinkage is 2 percent. The forward length is 18 percent less and the rearward length is 22 percent more. These distortions are getting large.

As we look at the chart we see that it is impossible for us to achieve the star Alpha Centauri. We are limited to our solar system. We can dream of traveling the stars. It makes good sci-fi. Doppler space time destroys those dreams.

If we do find any viable planets we can send robots to them. In my novel Futureoids, Gan and Raz took the seeds of man and woman to another planet which had chimp/ape type creatures. They impregnated the animals and then watched over the new humans until the human race was established.

We cannot bring ourselves to the stars but in the future we will have the technology to bring man there. Only those planets that have risen to a very higher level of scientific achievement and morality could populate the stars with their DNA patterns. It is not known if God will approve of such things but if God judges a planet by its ability to do such things then it will be done.

Chapter 22: The Neutron and the Hydrogen Atom

Section 22-0: Introduction

Quantum mechanics breaks apart the structure of the universe into subparticles which appear to explain the actions of the dot-waves as they appear in the laboratory under strong electromagnetic forces. Additional photons when added to the moving proton or electron produce interesting things which only last for short periods of time.

One explanation is that what we see and make in the lab are subparticles which are made of photonic energy and therefore present universe energy. The electron and proton themselves are quite sturdy and appear to be the product of very compressive forces such as the big bang or exploding black holes.

This means that stable subparticles must be the product of the three universes whereas unstable subparticles are merely the product of photons which are present energy. When we produce atomic bombs or hydrogen bombs, we are only dealing with photonic type energies.

The proton has a huge amount of energy but unlocking that energy is very difficult. Notice in the lab that we can produce both matter and antimatter although anti matter does not exist outside the lab. All this shows is that photons have the ability to produce present universe protons and anti-protons.

When we look at our universe we find that the basic structure of our universe is the proton, the electron, and the neutron. There are other strange things that can be produced during extreme space compressions but these three ingredients are our primary concern.

The question is what is a neutron? Is it a fancy quantum mechanical wave type structure or is it merely a different state of the hydrogen atom? One thing we know about the hydrogen atom is that it is stable. We have to add 13.8 electron volts to release the electron from the proton. Therefore the electron is bound to the proton within the hydrogen atom.

When we look at the neutron we find a strange reaction. The neutron when bound together by photonic energy within an atom such as helium is quite stable. Once we remove the binding photonic energy of the helium atom, we find that a neutron alone self destructs.

The reaction is that we get a hydrogen atom and a neutrino. The neutrino is merely short term photonic energy. This is strange in that if we add energy to a hydrogen atom it releases the electron. However the neutron has more energy than necessary to produce a simple hydrogen atom.

The reason for this is that the neutron is a product of high spherical compressive forces. The neutron is no different from a hydrogen atom. They are one and the same. This means that the neutron is a compressed hydrogen atom in which additional photonic energy has been added.

We can now calculate the amount of energy necessary to produce a neutron. Of course we know the answer to be 0.782 MEV. All we need to know is the velocity of the electron in the neutron's orbit. In addition we want to know the size of the neutron. This is not a fancy quantum mechanical solution. It is merely an Einsteinian solution. This serves me well because I only have the math abilities of simple algebra. When I was young was able to do more complex math but my mind has always enjoyed simple algebraic explanations for things.

The atomic physicists and scientists can look at the world in a very complex mathematical manner. The only problem is that we all need to understand models of the universe that we can readily see and understand. What good are pages of fancy math if the average bright person cannot understand it?

It is difficult for me to tell if Quantum mechanics is the way groups of dot-waves work or merely a mathematical expression of how the dot-waves work. Every atomic physicist is happy with the results but I have seen no explanation for the binding energy of the hydrogen atom or the excess energy of the neutron. These energies are clearly Einsteinian energies. Let us now look at the Einsteinian type calculations for the hydrogen atom.

SECTION 22-1: THE ELECTRON IN NEUTRON'S ORBIT

In this section let us review the hydrogen atom and then apply the results to the neutron. Within the hydrogen atom, the electron orbits the proton at the Bohr orbit and at a speed of $C/137$. This is the lowest orbit for the hydrogen atom. The radius of the Bohr orbit is:

$$R_{(Bohr)} = 0.529177E\text{-}10 \text{ meters} \tag{22-1}$$

In the hydrogen atom the negatively charged electron moves with a velocity of the speed of light divided by the inverse of the fine constant which is 137.036. If we add additional photon energy to the electron it will move to a higher orbit away from the proton. If it radiates energy from the higher orbit it will move to a lower orbit. Thus the electron captures and absorbs photons and emits photons. This gives us the light spectra.

Let us now look at the neutron. The neutron is merely the lowest state of the hydrogen atom. It is a hydrogen atom that is compressed spherically as with the big bang or a star implosion. In the past, energy was pumped into the neutron. If we take a hydrogen atom and add photon energy, in general we excite the atom to a higher state and free the electron. However, if we spherically pump the energy into the electron, we prevent its escape and force it toward the proton.

The neutron in free space tends to be unstable. A charged particle such as a proton tends to electrically match the structure of the universe that is charged dots of value Q at the radius of the universe. Neutrons do not match the universe. They will tend to break apart and form a swirl of dots. The electron is the mass for the swirl of charges. The neutron is a very simple device. It is an equal amount of plus and minus dots. The dots have both AC motions and DC motions. Patterns of moving dots produce binding energies and magnetic fields. Yet, at a distance the neutron is a perfectly balanced electrostatically but not magnetically.

Dots can have AC oscillatory motion. They can also form patterns of plus dots moving in one direction and minus dots moving in the opposite direction. These will produce magnetic fields perpendicular to the plane of motion. Thus the neutral neutron is not quite neutral.

Let us now investigate the structure of the neutron. Is the neutron a pure collection of dots with inner motions, or is the neutron a proton with a low flying electron in the neutron's orbit? The solution to this problem depends upon the energy calculations for the electron in the neutron orbit. It is possible for the electron to orbit the neutron if the energy of an electron in a low flying orbit is less than the total differential energy between the neutron and the hydrogen atom? The electron in this case would be a line of charge.

However, if the energy required to bring the electron near the proton uses up all the energy difference, then the electron will merely become part of the neutron. In this case, you will not find the electron in the Neutron's orbit.

Let us now calculate the amount of energy released when an orbiting electron from the neutron's orbit moves into a Bohr orbit and the neutron is changed into a hydrogen atom. Let us first look at a chart of the properties of the neutron, hydrogen atom, proton, and electron.

Structure	Mass in Kg	Mass in MEV
Neutron	1.67493E-27	939.565
Proton	1.67262E-27	938.272
Electron	0.910939E-31	0.510999
Hydrogen Atom	1.67353E-27	938.783

We see that when we move from a neutron to a hydrogen atom we lose 0.785 MEV. (Million electron volts) I have made this chart to six places from U.S. Government standards data. In addition, by adding the electron mass to the proton mass, I have prorated the mass of the Hydrogen atom.

Let us now calculate the energy released when the electron moves from the lowest Neutron orbit to the Bohr orbit. We know the answer is 0.782 MEV. For this analysis we will assume that the radius of the proton is reasonably equal to the wavelength of the proton. Thus:

$$R_P = 1.32142E\text{-}15 \qquad (22\text{-}2)$$

Using the same equivalency, the radius of the neutron would be reasonably the same as its wavelength. Thus:

$$R_N = 1.31959E\text{-}15 \qquad (22\text{-}3)$$

In equation 22-3 we state that the radius of a neutron is identical with its de Broglie wavelength. The neutron is slightly smaller than the proton. Thus:

$$R_N < R_P \qquad (22\text{-}4)$$

Let us now calculate the velocity of the electron in the neutron's orbit. This velocity will be a steady state velocity for the case where the electron stays in orbit. It will be a final transient velocity for the case where the electron merges into the proton.

The force acting upon the electron from the electrical world is:

$$F = K \, QQ/(R^2) \qquad (22\text{-}5)$$

This is the standard coulomb equation for the force between the proton and the electron. It is the same as used by Bohr. Let us now look at the centripetal force acting on the electron:

$$F = M \, (V^2)/R \qquad (22\text{-}6)$$

The opposing force is equal to the mass times the velocity squared divided by the distance. This is the same as Bohr calculated for the hydrogen atom.

Since the velocity of the electron is nearly the speed of light at the surface of the neutron, the mass of the electron is:

$M = M_E / (1 - (V/C)^2)^{1/2}$ (22-7)

This is Einstein's famous equation for the variation of mass and velocity. This equation must always be added to all Bohr orbit equations. It is necessary even at low speeds as will be seen at the end of this chapter.

Since the forces are equal we can equate both forces to each other.

$M_E (V^2) / R [1-(V/C)^2]^{1/2} = K (Q^2)/R^2$ (22-8)

Simplifying the equations we get:

$(V^2)/[1-(V/C)^2]^{1/2} = K Q^2 / M_E R$ (22-9)

Equation 22-9 provides us with the relationship between the electron's mass-radius (M_E R) and velocity (V) as it approaches the proton. If we know the velocity, we can solve for the mass-radius. If we know the mass-radius, we can solve for the velocity. However, this solution is more complex and it is best handled by a chart of velocities that match the distance in question.

The problem we have to solve involves the energy difference between the neutron and the hydrogen atom. The differential mass/energy is:

Differential Mass = $M_N - (M_P + M_E)$ (22-10)

Since M_N= 939.565MEV, M_P=938.272MEV, and M_E= 0.510999MEV, we get:

Differential Mass/Energy = 0.782001MEV   (22-11)

The Einsteinian mass increase in MEV of Equation 22-11 for the electron must be added to the original rest mass of the electron to produce the total mass.

$M_{E(Total)}$ = M_{Eo} + Differential Mass   (22-12)

$M_{E(Total)}$ = 1.29300MEV   (22-13)

The ratio of the total mass of the electron in the Neutron orbit or just before being absorbed by the neutron to the rest mass of the electron is:

$M_{E(Total)}$ / M_{Eo} = 2.530338   (22-14)

In equation 22-14, we see that the electron in the Neutron orbit or just before absorption has a ratio of the Einsteinian mass to the rest mass of approximately 2.53

times. This means that the electron is traveling at close to 91 percent of light speed. We can now use equation 22-9 to solve for the velocity and the radius. The velocity is easily found from Einstein's formula, equation 22-7. Then we can find the distance from equation 22-9. The alternate way of solution is to prepare a chart. This helps to see what is happening.

We can rearrange equation 22-9 as:

$$R = [K \, Q^2 / M_E \, C^2] \cdot [1 - (V/C)^2]^{1/2} / (V/C)^2 \qquad (22\text{-}15)$$

The first part of equation 22-15 is:

$$K \, Q^2 / M_E \, C^2 = 2.817961E\text{-}15 \qquad (22\text{-}16)$$

We can now make a chart of the electron velocity verses the radius and mass of the electron near the surface of the proton:

V/C	$[[1-(V/C)^2]^{1/2}]/(V/C)^2$	Radius	Mass
0.919	0.4668195	1.31548	2.5364
0.9188	0.467575	1.31760	2.53342
0.9186	0.4683295	1.3195698	2.53043

In the chart we find that the mass of the electron matches the differential mass of the neutron when the velocity reaches 0.9186 C and the radius is the same as the radius of the neutron's wavelength. The chart tells us that the electron does not maintain an orbit around the proton but merges into it because this distance is less than the radius of the proton.

The mass that was necessary to merge the hydrogen atom into a neutron is only the Einsteinian mass. The electron is moving at 91.86 percent of light speed as it merges into the proton. Of course, this must be forced by spherical compression. In general ordinary added energy pushes the electron far from the proton.

SECTION 22-2: THE BOHR ORBIT

Let us calculate the velocity of the electron in the first Bohr orbit state. Bohr did an excellent job of bringing an understanding of the hydrogen atom to man. He compared the light spectra and simple equations to relate the electric field and the centripetal force. He derived a set of equations, which fit into each other. He did not explain the binding energy of the hydrogen atom. He merely used the 13.58 electron volts in his work.

Let us now calculate the Bohr orbit without any need to study the light spectra. Bohr studied the spectra and came to his great conclusions. However this method of empirical analysis which fits the data into the equations lacks the understanding of what is happening in the process. This chapter will show the reader the exact workings of the hydrogen atom and how it was produced from the neutron.

In section 22-1, The Electron in Neutron's Orbit, it was shown that the additional mass of the neutron was due to the electron in a final orbit of radius 1.3197E-15. This produced a gravitational mass of 2.53043 times the rest mass of the electron. The characteristics are as follows:

$$M_E = 2.53043\ M_{Eo} \qquad\qquad\qquad (22\text{-}17)$$

$$R_E = R_N = 1.31957E\text{-}15 \qquad\qquad (22\text{-}18)$$

$$V_E = 0.9186C \qquad\qquad\qquad\qquad (22\text{-}19)$$

In equations 22-17, 18, and 19 we see that the mass of the electron in the final neutron orbit is about 2.5 times the rest mass of the electron. The velocity was then calculated using the standard equality between the centripetal force and the coulomb attraction. Thus:

$$KQQ/R_B^2\ = M_{Eo}\ (V^2)\ /\ R_B\ [1\text{-}V/C)^2]^{1/2} \qquad (22\text{-}20)$$

In equation 22-20 we find that the electric force equals the centripetal force when corrected for the gravitational mass increase as per Einstein's formula.

For the electron in the neutron's orbit, the Einsteinian orbital gravitational mass calculated by Equation 22-13 is:

$$M_{Eg} = 2.53043\ M_E \qquad (1.293\ MEV) \qquad (22\text{-}21)$$

The Einsteinian mass increase from equation 22-21 is the difference between the mass of the electron at the orbital speed and the rest mass of the electron. Thus:

$$\text{Delta Mass} = M_{Eg} - M_{Eo} = M_{Eo}\ /(\ [\ 1\text{-}(V/C)^2]^{\ 1/2}\ ) - M_{Eo}\ (22\text{-}22)$$

This calculates to be:

Delta Mass = 1.53043 M_E (0.782MEV)   (22-23)

In equation 22-23 the gravitational mass, which gives the neutron extra, weight has a surplus energy of 0.782 MEV.

Together there is sufficient energy to radiate photons and neutrinos and to produce the Bohr orbit. The equations of the Bohr Orbit and the Neutron are complicated by the Einsteinian mass increase. Notice at low speeds such as the first Bohr orbit with V= C/137 and higher orbits, the value of the binding energy will be the same value as the Einsteinian mass increase.   Thus we can look at the Einsteinian mass increase and use that for the Binding energy.

Let us now look at the Bohr orbit. The neutron will radiate its surplus energy and the velocity of the electron will move from relativistic speeds to ordinary speeds. Bohr deduced that the first orbit would be at C/137 and R_B at 0.529177E-10. (This is a modern number from U.S. Government data). He showed that there was a constant relationship between the square of the velocity of the orbit and the radius such that:

$M_E(V_B{}^2)R_B$ = Constant   (22-24)

For an orbit of the series R_B, $4R_B$, $9R_B$... the velocities would be V_B, $V_B/2$, $V_B/3$. The proton electric field attracts the electron and the electron heads toward the proton. Bohr had no explanation as to why the electron would not be captured by the proton. Yet his basic answers were great.

As the electron moves toward the proton, its Einsteinian mass increases, which causes additional gravitational repulsive forces away from the proton. In addition, the proton transfers some energy to the electron that increases the mass of the electron and decreases the mass of the proton. This helps stabilize the electron in the lowest Bohr Orbit. Additional magnetic attractive binding forces due to the synchronous motion of the electron's AC field with the proton's AC field accompany this. The fields are not perfect AC fields. They are modulated DC fields that do not form perfect sine waves. Harmonic currents, which are synchronized, form attractive fields. It is complex but the fields are attractive.

Let us now look a little more carefully at the electron in the Bohr orbit at V= C/137 to see if 137 is the correct answer or if some other answer is correct. Let us use Einstein's formula:

Correction = $[1-(V/C)^{1/2}]$ (22-25)

Let us now look at the differential mass as we change the velocity around C/137. The following chart shows the calculations.

Velocity	Correction	Error	Energy (EV)
C/274	0.9999933	6.56PPM	3.4 EV
C/200	0.9999874	12.5PPM	6.387EV
C/137	0.999732	26.6PPM	13.59EV
C/100	0.999499	50.0PPM	25.55EV
C/10	0.9949	5012PPM	2561EV

From the Chart we see that a velocity of C/10 produces a relativity mass error of 5012 parts per million, (PPM) or 2561 electron volts, (EV). Even at C/274 we get a small error in mass due to Einstein's equation. At C/137 we see that the differential relativistic mass has an error of 26.6 PPM or 13.59 electron volts. Now we know where the binding energy of the hydrogen atom comes from. It is identical to the gain in relativistic gravitational mass as per Einstein's formula. However, it comes from the transfer of mass from the proton to the electron that is identical with the above numbers at the Bohr velocities. The above chart is the proof that the binding energy of the electron is identical to the electrons mass increase with velocity as per Einstein's formula. As the electron moves faster and faster in the Bohr Orbit, it pulls energy from the proton to it.

Although the proton has a particular mass it contains a certain amount of free photonic energy. Therefore there is a basic proton ground state and many higher levels of energies. Therefore there are no two protons in the universe that contain the same exact amount of dot-waves within their structure.

The same is true of the neutron and all the subparticles. Since everything contains a degree of photonic energy we can only say that on the average the proton mass is what we measure.

In any event this chapter shows that Einstein orbital space time gives the correct answers for the binding energy of the hydrogen atom and the surplus energy of the neutron.

It appears that all the theories are imperfect. The best we can say is that they match the experimental data to within a certain error band. There is always random chance and chaos in the universe and this reflects within our measurements.

Chapter 23: The Triple Universe

The equations of the dot-wave theory look at the present universe as being a product of a big bang 13.7 billion years ago. This time is due to astronomical measurements from the Hubble telescope and based upon astronomers methods. The dot wave analysis comes to the same conclusions from the matching of dot-wave equations and the proper constants.

The general solution to the universe is a solution in which the light speed can vary from zero to infinity. There is a period of time of non-linearity after the big bang in which the external pressure from God's spiritual energy produced the initial creation. Once the universe was stabilized and the first Earth was created in multiplicity all over the universe, we entered a period of linearity.

This linear period caused the universe to look like an exponential function which expands more rapidly with time. Within our dimension an increasing light speed and an increasing ruler causes our measurements to appear that the light speed is constant. There are always non-linearity's and astronomers today look at the universe as expanding faster than the speed of light. In truth it is common mode which we could not detect readily. It would be the non-linearity's that give the appearance of us moving faster than the speed of light.

These measurements clearly show that the present universe is erasing and that energy is transferring into another time dimension. This indicates that we live in a sandwich of three universes which are the past universe, the present universe, and the future universe.

As we try to understand how the mind of God works, it becomes self evident that from God's point of view the past, the present, and the future all coexist simultaneously. That is the way the mind of God works. Everything must flow in a circle.

The beginning of God's creation must be the end as well. There may be many steps in between but the end must always be the same as the beginning. Any structure that God designs must start with God and return to God.

When we consider the multi light speed dot-wave universe, two equations are important.

$$Energy = M\ CC \tag{23-1}$$

$$Wavelength = h/MC \tag{23-2}$$

For the case where we have an equal energy distribution of dot waves within a particular spectrum of light speeds, equation (23-1) shows us that the mass of the dot-waves at the very lowest light speeds is very high. When we substitute equation (23-1) into equation (23-2) we get:

$$Wavelength = hC/Energy \tag{23-3}$$

Equation (23-3) specifies that as the light speed increases, the wavelength of the dot-waves also increases. As the universe moves upward in steps to higher and higher light speeds, the size of the universe reaches toward infinity.

Our present universe operates at what appears to be a constant light speed. Our prior universe operated at a slightly lower light speed and our future universe operates at slightly higher light speed. The differences between these light speeds are very tiny. It is a small step. When we look the distances involved between the universes they are less than the size of a proton. The size of a dot is almost zero and the difference between universes is in the range of the size of a dot.

If we look at our prior universe which is slowly erasing, we find that the stars and planets explode. A time is reached where the protons and electrons disintegrate. This causes the structure of the prior universe to fall apart and turn into chaos. Chaos is a situation in which the dot-waves do not form meaningful patterns.

In many respects the situation is similar to the general gas law. When gasses have some frozen particles within them they tend to occupy a smaller volume. As the gasses evaporate, they expand and exert pressure upon the surface of the container. If the container is a balloon, the balloon will expand.

The prior dot-waves interact with us. Our universe is like the surface of a balloon. The net result is that the destruction of the prior universe causes its structure to turn into chaos and then expand. The result is a uniform expansion which tends to push galaxies apart. Therefore the source of the dark energy is the destruction of the past universe which converts protons and electrons into raw dot-waves in chaos.

As we look at the future universe we find that black holes which formed within the past universe penetrated into the future universe. The universes overlap by tiny distances. When a black hole explodes it will cause nonlinearities in our space time. Since the past, the present, and the future all coexist, the forming future universe will affect the present universe and the past universe as well.

The process is continuous. Old galaxies are slowly destroyed, and new galaxies are continuously formed. The universe is a product of an original big bang and a continuous creation process. To the astronomer it looks like a simple single light speed universe. However it is a variable light speed sandwich of three universes all tied to each other.

Someday the electrons and protons of our galaxies will break apart and explode. This destruction can be called a little bang as it tends to be mild as compared with a big bang explosion of a newly forming galaxy.

The formation of galactic structures across the time barrier to future galaxies causes a back pressure upon the present galaxies. This is similar to an additional gravitational force which is similar to additional mass operating upon each present galaxy. The dark matter that is necessary to hold the galaxies together is due to the energy from the prior galaxy which flowed into the future and formed new galaxies.

The driving force in the system is God's higher light speed energy. This will force the energy of the dot-waves to move upward in light speed as each step of the universe occurs. It is God's higher light speed spiritual energy which contains the universe. God is the driving function which compresses the universe during the initial creation and which releases the compressive forces slowly over huge amounts of time.

The exact mathematics of the three universes will involve some very complex math that few people could understand. The concepts are simple. Instead of a light sphere moving from an explosion at a single point at our light speed, there is an initial explosion and then a spectrum of explosions.

The first explosion is at a very low light speed. One black hole produced a huge amount of galaxies with billions of individual black holes. The energy from the black holes flowed into the second next galaxy and at a somewhat higher light speed.

The process continues so that instead of a linear motion of the surface of the universe, we got a series of steps. When we look back in time we find the results of many steps and transformations. Instead of an analogue universe of linear movement, we get a digital quantized universe. If all the little steps are small enough, it will be very difficult to tell if a galaxy that we look at is an original galaxy or merely a recreated copy of the same galaxy leading back in time to the original galaxy during the creation process right after the first big bang.

Although the light speeds of the three universes are continually rising, they are always one step apart from each other. Therefore the three universes operate at a differential light speed. If you move into a future universe this universe will be the past and an additional forming universe will be the new future.

The question is how many years apart are the universes in development? The initial creation most likely caused future galaxies to be exact copies of prior galaxies. The process was homogeneous at the beginning. Then during the continuous creation period, new galaxies would form out of old galaxies in a heterogeneous process. We know that we are 13.7 billion years from the big bang. We also know that Our Earth is several billion years old. Therefore there is a window of life in the universe and the development time between universes must overlap the development time. This will insure the availability of future planets for the transformation of souls.

Although the souls of man could be digitally stored for billions of years, it appears most likely that the higher Earths exist today. Therefore the development times overlap between the universes.

The energy of a black hole flowing into the second future universe and then exploding at the same location would causes space time distortions but would not be very evident to our astronomers. At the same time spiritual interactions between the past and future galaxy would tend to copy it exactly.

On the other hand, there are vast areas of space time which appear quite empty to us. It is these areas that must have contained galaxies from the far past. It is possible that new galaxies have formed in these areas and are already being copied upon a second future universe. We can say that there are three universes but we can also say that there can be a whole spectrum of time dimensions. To us we see three universes but to the mind of God things could even be more complicated.

God designed the universe so that crude original existence will continuously move up in light speed toward higher and Godlier quality. Every time a universe erases and reforms, the prior energy turns into chaos. Whatever existed at the beginning of the process no longer exists.

All we see and measure are copies of what used to exist. As we look at our universe we see that many things are being destroyed. Eventually everything that we see here will be gone. However as we move into the future galaxy we will find that what has been destroyed often has been reborn. In effect galaxies reincarnate. They give birth to themselves. There is always some random chance and chaos in the process. Yet for the most part, this Earth will be born again.

The upward motion of the Earth to higher light speeds will cause the quality of the Earth to improve. The bodies of man will be improved as well. The reincarnation process will insure that the moral quality of man will improve as well.

Chapter 24: The Speed of the Dot-waves

The dot-wave equations are written from the point of view that the dot-waves travel at our light speed within our universe. In reality combinations of the dots themselves can travel from zero light speed up to light speed infinity. What we see and measure is not the speed of the dot-waves themselves but the speed of huge numbers of dot-waves which combine to produce photons and particles.

When the original universe was compressed to a near pinpoint, the radius of the universe was very small and the light speed was near zero. The light speed equation is:

$$LS = K\,Ru \qquad\qquad (24\text{-}1)$$

Equation (24-1) specifies that the light speed is proportional to the radius of the Universe. All our scientific measurements and experiments indicate that our light speed is constant. This is true because the universe is quantized with respect to light speed. There is a band of distances from the center of the universe outward in which the light speed is constant.

As the next universe forms, the light speed structures of the dots moves upward. The new future universe is formed out of the energy of the past universe and the image of the present universe. Some of the image of the past universe still exists while the future universe is being formed. This is especially true when the future universe starts to form initially. This produces a strong image and insures that a basic galaxy will repeat itself over and over again.

The initial big bang was a singular event. The second universe formed out of the first but it was necessary for the mind and body of God to form it. Therefore God acted on both the original universe and the second universe. In so doing, the third universe was assured.

In the laboratory we measure the light speed of groups of dot-waves with basic infinite light speed capability but which are part of patterns within the radius of our universe which astronomers calculate at 13.7 billion light years. This has been confirmed by the dot-wave calculations of this book.

The future universe is somewhat larger and the light speed is somewhat higher. How many levels of universes there were before the present is unknown and how many levels of universes are after the present is also unknown. All this information can only come from God. All we can do is see and measure a very small part of the entire creation. However the number is not really important since the light speed jump steps are very tiny as well.

We only see and measure the effects of the dot-waves. The cause of the structure of our present level of the universe is the dot-waves. We cannot measure the cause which exists in the mind of God. We can only measure the effect.

In many respects the entire universe is similar to a hydrogen atom. There are quantized steps in the levels of the atoms. When the atom is in a low energy state, the size of the atom is smaller and the electron is within a lower shell or lower radius from the proton. As energy is added to the electron, it moves into a higher state. Of course this is the Bohr model of the atom. Mathematicians will produce other models with complex equations. However the simple Bohr model works quite well.

The universe itself operates in the same manner. From Einstein's equation we get:

$$\text{Energy} = M\,C\,C \tag{24-2}$$

In this equation if we convert all the energy into a huge mass at near zero light speed, we find that for these conditions the energy is at a very low state. Thus when

God compressed the universe toward a pinpoint the energy was converted into compressed energy or matter.

For our level of the universe this is not true. When we compress photonic energy to produce mass we do not reach a lower energy level but we merely transform energy into mass. Yet the entire universe itself operates differently.

The initial universe was in a low energy high mass state at near zero light speed. This is an unstable situation. It was the pressure exerted by God which maintained the ball of mass. Once God released the ball of mass, it exploded thereby converting mass to energy. God then contained the explosion to a fixed radius. After a period of creation, God released the fixed radius and this started the transformation of some of the mass/energy of the first universe into the second universe. Thereafter God set up the bands of radius leading up to the present radius.

Our universe operates within a band. Each new band operates at a higher light speed. The dot-waves are formed out of the energy of the prior universes. God's dot-waves coexist with the physical universes and control them. As our universe expands and erases, the energy of our universe decreases. In addition the patterns of the dot-waves drop and turn into chaos.

In the future upon our level of universe, there will be no structures left. In addition the number of dot-waves will decrease to a minimum. As we look at the Bohr atom, when an electron moves to a higher shell, the basic structure of the lower shells remains. The electron can return.

In the same respect the basic shells of the entire universe remain. There is a certain amount of raw energy within the prior universes. This aids the production of a new creation far into the future.

It is interesting to note that the Bohr atom operates in a digital manner and that the entire universe also operates in a digital manner. The entire universe is digitized as well. As we move upward to a higher shell of the universe we move away from the material state. Notice that at a certain energy level the electron escapes from the atom. The same is true of the physical universe. At a certain energy level or distance from the center of the universe, our physical shell will escape physical existence.

At that time only spiritual existence in the mind of God is possible. Some people hope for future spiritual existence. Such things are certainly possible. It appears to me that we must return to a state of God alone. However there may be a long period of time when collectives of individuals remain.

Many things are possible. This all depends upon God. At any time God could reproduce a soul which has had many lives and became part of a collective soul. The individual could be pulled out of God's mind to experience the awe and wonder of the entire experience.

It gives me comfort to believe that I only have a limited number of lives to experience. Ultimately God is God and we are little creatures within the mind of God. Our fate is not in our hands but God alone.

The dot-wave theory is a work in progress. There is much more work to be done. There are many different ways at looking at the dot-wave. It could be looked at as a spherical expanding ball focused upon a single point at a distance Ru from the outer shell. It could also be viewed as a planar current loop of radius Ru. This produces a point magnetic field at the center of the loop and pointing perpendicular to the loop. If we add all the point magnetic fields together with part of the circumferences meeting at a common center point, we produce a universe with of a surface sphere at a distance of Ru from the center.

Of course this looks at the universe from a purely electrical analogy. If we look at the universe from complex wave equations we will get different answers. It may very well be that the electrical analogy or model of the universe is just an engineering method of describing a very complex mathematical universe.

In truth the dot-wave theory may only be an equivalent of the universe that an Electrical Engineer could understand while the true solution is much too complicated so that only God could understand such things.

It may very well be that God chose me for this task because my ability to produce an equivalent model is the limit man can do to understand God and the Universe. The actual structure of God and the Universe may very well be so far above us that no one can every really describe it.

In any event, the dot-wave theory is a method of understanding God and the universe from a simple Engineering perspective. It could be said that it is a best fit approximation to the more complex reality of our existence.

Chapter 25: Conclusions

This book has been a part of a thirty three year study of God and the Universe from the perspective of how God would view the universe. It is an attempt to put myself and the reader into a different frame of reference.

Instead of looking upward from our mind toward God's mind, we look downward from God's mind to us. This is a very difficult task. In general we look outward to the world through our eyes. When we try to look inward toward God we must first find our spiritual dimension.

Once we achieve communication with our spiritual dimension then we start to look into the mind of God. The ability to look inside ourselves gives us a glimpse of what lies beyond our existence. Some people will experience various dreams and visions which provide information about the world beyond.

The study of the spiritual information over many years gives us an understanding of what lies beyond our existence. Most people are merely concerned about their own salvation and the information they receive is satisfactory for that purpose. Some of us want to know more than that. We want to know if God exists and what God is.

To know such things gives us a little peek into the mind of God. We are little finite creatures attempting to understand a God of infinite intelligence and power. The study can be very frightening because we have no ability to defend ourselves against this powerful entity we call God. Somehow we have to overcome our fear and look God in the eyes.

When we do this we do not find a horrible entity. We find a very moral and ethical entity. God is not human. God is eternal space time intelligence. Some people like to put human traits into God. They conceive of a God who would punish an individual man for all eternity. This is the result of pure fear. The God of the Universe is not that kind of God. In addition everything that God creates must in the end perish. In the end of the process God must always return to the original state of God.

All the creations of God must return to the state of chaos. There is no suffering in chaos. There is no existence in chaos. There is no foundation for the concept of eternal damnation for anyone. In the same light there is no foundation for the concept of eternal existence for anyone. Only God is eternal. Everything else is temporary.

The life we have is what we see. Our soul lives on and many of us will live again. We will remember very little of our past lives. For those who travel to higher Earths the purification process will eliminate this existence as if it never occurred.

When we travel from one Earth at this level to another, the speed involved in the transmission is 1000 light years per second. We can travel to the center of our galaxy in just 30 seconds. In order to travel to the higher universe, we must pass through the center of our galaxy and move into the center of the future galaxy in the next universe.

Just as one galaxy will give birth to itself in a series of universes, those who achieve the higher Earths will be duplicated in a similar process. The means of the duplication is through our souls. In the same light a galaxy has a soul. There is a spiritual interaction between the energy of a galaxy and the spiritual energy of God.

This interaction causes a galaxy to reproduce itself. We then get complex physical/spiritual interactions for the reproduction of galaxies and Earths. Our Earth already exists across the time barrier. Some things do change. The dinosaurs were part of an original Darwinian type process. They were necessary for the production of life from

216

bacteria. They are merely part of an original creation in which the hand of God took direct control.

Once God set the ball rolling, the duplication process took over. The triple universe insures that galaxies will be duplicated and repeated for a long time to come. The dinosaurs are not necessary or desired upon the higher Earth and beyond. They served the purpose of clawing their way into existence. They upgraded the God of the Earth to higher intelligence leading to the final creation of man.

Once we go beyond this Earth, only reincarnates will appear. Simple creatures will be the parent of the human race. As we continue upward on our journey to the far future, things will change. The many galaxies will fade into fewer and fewer galaxies. The populations of man will combine and drop. In the end of the process a very long time from now, only one Earth will appear. Finally that too will be gone. All that will be left is a spiritual Earth in God's energy.

God took a spiritual Earth and recreated it in the physical world. Once this was done, pure good became a mixture of good and evil. However puppets of God became real for awhile. We are real. We are independent creations of God because we were created out of chaos. We have limited free will since in the end we must return to a robotic state. We must return to a controlled robot state within the mind of God.

Since we came from chaos, the process merely returns the majority of people back into chaos. For most people there is neither punishment nor reward. There is only temporary existence and in the end non-existence. In many respects non-existence is very peaceful. There is no pain or suffering or thought. Man started as nothing and in the end man will return to nothing.

Although many of us must go on to serve God again and again, it is quite comforting to realize that someday we will become nothing at all. The thought of being a God is quite horrible. The thought of living forever is quite horrible. The thought of suffering a limited number of lives is okay. This is especially true if one moves upward to higher and better existence.

On the other hand, at this level of existence we have maximum freedom of spirit. We are all slaves in one way or another. Many of us are economic slaves and live hand to mouth. Many of us go to bed hungry at night. There is a distribution of people with greater and less degrees of suffering.

Some of us are lucky to have enough food and decent shelter. To be able to be free of work and to be able to contemplate God and the Universe is a luxury that few can afford. Often God provides for those who serve God in one way or another.

I have worked this task for thirty three years. It has only been a part time job. There are periods of dreams that compel me to study new ideas. In my sleep data from my spiritual mind enters my physical mind. Then I study the new data. Some years go by without any new thoughts. Then there will be an active period where the thoughts flow into mind every night. In the morning I type them out. Then I can study them and look for new possibilities.

It is an interesting task in that my mind is filled with ideas as I sleep. I have been retired since 1993 at age 55. This has given me the freedom to do this job when the dreams compel me to write. At this present time I am getting a feeling that my understanding of God and the Universe has reached a high level. Over the years I have never felt for long that my work was adequate. Now I am beginning to see the big picture. Surely I could work on this for many more years. Yet my time clock in running out. Soon I will have to retire from this task. I have done the best I can for God.

One thing that gives me a degree of happiness is equation 19-48. From 1981 to 1983 I struggled to find any constants that would give the expression any meaning. The astronomer's data I had at that time was different than the present numbers. I could not find a match in my numerical analysis. I do not have the ability to do complex math as Einstein and other great mathematical physicists could. Yet I knew full well from my dreams and visions that the universe was put together with simple algebraic expressions.

I did not intend to correct my dot-wave theory which I published in 2000. Then as this book was sent to CreateSpace for the initial set up, there was a television program on the Hubble telescope. Suddenly 13.7 billion years was the accurate time from the Hubble data. I then decided to change my calculations using the Hubble data. It was quite shocking to me to discover that a vector type correction of 0.866 was the missing link and that my calculations match to within a very small difference.

Now that I have this information I can continue my work on the dot-wave theory. A few months ago I though I no longer had any purpose for my note books. Futureoids was to be my last effort. I threw everything away into the ash can. In my small house I have very little room for things. I used to always have nice size houses with plenty of room. Yet I no longer own a house and I rent a small house from my older daughter on her husband's family farm land.

Over the years I have worked part time as a handyman. I like fixing things and building things. During the last year I rebuilt this small house. It now has modern bathrooms and is small but comfortable. I used to have only a small plot of land to live on. Now I have plenty of land to roam on. Across the street I have 400 acres of private farm land. It is a different way of living.

Which is better, a large house on small land or a small house on big land? Perhaps they are the same.

I never expected to return to the dot-wave theory. The study of God has always been emotionally draining. To contemplate life and death is not fun. Yet I was forced to do this. I tried to escape from this task but I was unable to do that. The only pleasure in this work was the dot-wave theory and Doppler Space Time.

This book is at an end. My writing may be over but I will continue to study and revise the dot-wave theory. I still have a lot of work to do on this little house. I have trees and flowers to plant as well. The sandy soil seems to make the garden grow better. In the past I had a lot of clay type soil and carrots would only grow one inch high.

Perhaps now I can now retire from this effort and watch the carrots grow in the garden. Perhaps that is the only thing really important in our lives.

Index

Alpha Centauri: 191
Angelic salvation: 11, 31,2, 66, 81
Angels of god: 11, 32, 43,5, 74,5, 221
Atheists: 9, 62
Bacteria: 10,4,9, 28, 36, 54, 60,3,4,8, 82, 217
Chimp/Ape: 19, 20,8, 77
Communicating with the earth god: 80-83
Concepts of dot-wave theory: 99-105
Conversion of mass to charge: 134-152
Cosmic reincarnation: 34, 44, 60,7,9, 70,3,6, 84
Creation: 3,6,7, 13-7, 20, 32,7,9, 45-9, 52-6, 73,5,7, 82,7, 103, 192, 205-8, 211-12, 216-7
Creator: 3-5,8, 52, 73, 221
Dark energy: 88, 116, 206
Dark matter: 88, 207
Darwinian: 14,9, 20, 62,3, 216
Data flow from soul: 33-39
Destruction: 10, 31, 47, 71,67, 82, 160, 206
Dinosaurs: 28, 216
DNA: 7, 20, 77-9
Doppler space time: 17, 23, 84,5, 98, 104,6, 152-4, 220
Dot-wave theory: 1, 72-4, 88, 90-9, 101-109, 120, 129-33, 146, 204, 213-4, 219, 220
Dreams: 34, 41, 53,5,9, 73, 96, 183, 215, 218, 219
Earth God: 7, 12,3,6, 29, 31,6,7, 41-4, 51-8, 63,6, 73, 76-9, 80-83, 96, 221
Einstein: 1, 16-8, 22-3, 84-6, 90-4, 106,8, 128, 133, 146, 154-181, 184, 190,1, 194, 198-203, 211, 219
Electron: 16, 85-8, 107, 111, 117-9, 120-2, 128-9, 143, 147,9, 149-52, 155, 159-60, 182-3, 194-203, 207-7,212-3
Elephant: 64
Engineer: 1, 17-8, 22-3, 84, 90-98, 103-9, 133-4, 148, 152,4, 188, 196-7, 214, 221
Eternal God: 1
Eternal suffering: 68
Evil: 7, 32,5,6,8, 41-5, 54-60, 74, 92-3, 217
Evolution: 14, 28, 62-3, 77
Fetus: 20, 49, 76, 79
Future universe: 5,11,83-8, 90-3, 101-3, 183, 204-11, 222
General solution: 154, 204
Gods of man: 4, 8-15, 32, 40, 57, 79

Standard Value MKS system

R_{bohr} = 5.29177E-11
C= 2.99792E8
(Fine constant)$^{-1}$ = 137.036
G= 6.67260E-11
ε_0 = 8.85418E-12
h = 6.6208E-34
K = 8.98756E9
Mn = 1.76493E-27
Mp = 1.67262E-27
Me = 0.910939E-30
Q = 1.60218E-19
U_0 = 1.25664E-6
Z_0 = 376.731
Year = 365.242 days

Formula for Gravitational Constant

G = 16 pi e Uo / $(137.036)^3$ = 6.67223E-11

Formula for expansion velocity of Bohr orbit

GM_HM_H = 2Uo (QC/137.036) (4 pi Q V_B*) Cos 30°

V_B* = 1.21667E-28 meters per second

Tu = 13.7827 billion years

Formula for time of universe from Bohr orbit

4 pi Q Tu = Cosine 30°

Tu = 13.6306 billion years

Time of Universe from geometric mean of both solutions

Tu = 13.7064 billion years

Astronomical measurement

Tu = 13.7 billion years

Page for notes

www.ingramcontent.com/pod-product-compliance
Lightning Source LLC
Chambersburg PA
CBHW051901170526
45168CB00001B/190